PSpice for Digital Communications Engineering

PSpice for Digital Communications Engineering

Paul Tobin

www.morganclaypool.com

ISBN: 1598291629 paperback
ISBN: 9781598291629 paperback

ISBN: 1598291637 ebook
ISBN: 9781598291636 ebook

DOI 10.2200/S00072ED1V01Y200612DCS010

A Publication in the Morgan & Claypool Publishers series
SYNTHESIS LECTURES ON DIGITAL CIRCUITS AND SYSTEMS #10

Lecture #10
Series Editor: Mitchell A. Thornton, Southern Methodist University

Library of Congress Cataloging-in-Publication Data

Series ISSN: 1932-3166 print
Series ISSN: 1932-3174 electronic

First Edition

10 9 8 7 6 5 4 3 2 1

PSpice for Digital Communications Engineering

Paul Tobin
School of Electronic and Communications Engineering
Dublin Institute of Technology
Ireland

SYNTHESIS LECTURES ON DIGITAL CIRCUITS AND SYSTEMS #10

MORGAN & CLAYPOOL PUBLISHERS

ABSTRACT

PSpice for Digital Communications Engineering shows how to simulate digital communication systems and modulation methods using the very powerful Cadence Orcad PSpice version 10.5 suite of software programs. Fourier series and Fourier transform are applied to signals to set the ground work for the modulation techniques introduced in later chapters. Various baseband signals, including duo-binary baseband signaling, are generated and the spectra are examined to detail the unsuitability of these signals for accessing the public switched network. Pulse code modulation and time-division multiplexing circuits are examined and simulated where sampling and quantization noise topics are discussed. We construct a single-channel PCM system from transmission to receiver i.e. end-to-end, and import real speech signals to examine the problems associated with aliasing, sample and hold.

Companding is addressed here and we look at the A and mu law characteristics for achieving better signal to quantization noise ratios. Several types of delta modulators are examined and also the concept of time division multiplexing is considered. Multi-level signaling techniques such as QPSK and QAM are analyzed and simulated and 'home-made meters', such as scatter and eye meters, are used to assess the performance of these modulation systems in the presence of noise. The raised-cosine family of filters for shaping data before transmission is examined in depth where bandwidth efficiency and channel capacity is discussed. We plot several graphs in Probe to compare the efficiency of these systems. Direct spread spectrum is the last topic to be examined and simulated to show the advantages of spreading the signal over a wide bandwidth and giving good signal security at the same time.

KEYWORDS

Fourier series and Fourier transforms, baseband and passband modulation, pulse code modulation, time-division multiplexing, quantization noise, M-ary signaling, QPSK, QAM, eyemeter, scatter diagrams, spread spectrum, raised cosine filter.

I would like to dedicate this book to my wife and friend, Marie and sons Lee, Roy, Scott and Keith and my parents (Eddie and Roseanne), sisters, Sylvia, Madeleine, Jean, and brother, Ted.

Contents

Preface

Before each simulation session, it is necessary to create a project file as shown in Figure 1. Select the small folded white sheet icon at the top left hand corner of the display.

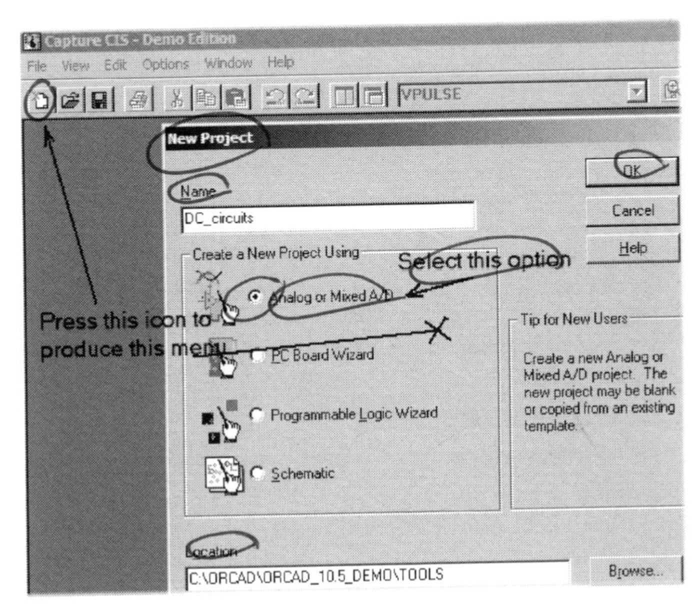

FIGURE 1: Creating new project file.

Enter a suitable name in the **Name** box and select **Analog or Mixed A/D** and specify a **Location** for the file. Press **OK** and a further menu will appear so tick **Create a blank project** as shown in Figure 2.

This produces an empty schematic area called **Page 1** where component are placed. Libraries have to be added, (**Add library**) by selecting the little AND symbol in the right toolbar icons. The easiest method is to select all the libraries. However, if you select **Create based upon an existing project**, then all previously used libraries associated with that project will be loaded. Chapter 1 uses the Fourier series expansion and Fourier transform to show the relationship between pulse width and pulse period by examining the spectra for different pulse signals. In chapter 2 we generate baseband signals and again examine the spectra for these signals. Chapter 3 examines another baseband modulation technique – the important topic called pulse code modulation (PCM). In this chapter we examine sampling, anti-aliasing

FIGURE 2: Create a blank project.

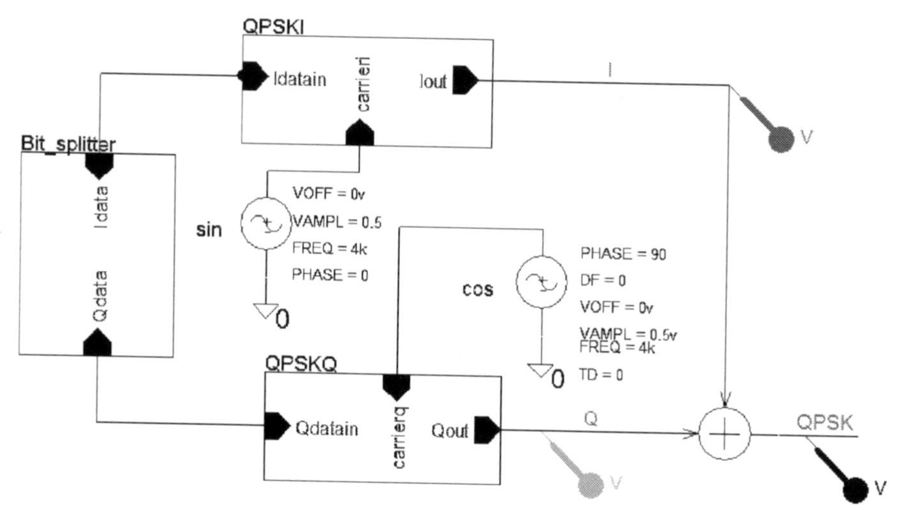

FIGURE 3: QPSK modulator hierarchical system.

filters, quantization noise and sample and hold. Also investigated in this chapter is time division multiplexing and we construct a single channel PCM system from transmission to receiver i.e. end-to-end. Passband systems are considered in chapter 4 where systems such as frequency shift keying (FSK), amplitude shift keying (ASK) and phase shift keying (PSK) and differential forms are considered. Chapter 5 considers multi-level systems (M-ary) such as quadrature phase shift keying (QPSK) and quadrature amplitude modulation (QAM). The hierarchical method of construction is used in these systems because of the system complexity. Figure 3 shows a QPSK modulator system broken into manageable blocks. The main schematic is named Figure 5-005 and the sub-circuits in different pages are named Figure 5-005a to Figure 5-005g.

Chapter 6 looks at systems performance where we introduce home-made PSpice meters for producing eye and scatter diagrams to assess the performance of the modulation techniques in the presence of noise deliberately introduced. The raised-cosine filter family, the integrate-and-dump filters, and zero forcing equalizers are also investigated in this chapter. The power of the macro is introduced to plot statistical probability curves and a bit error rate (BER) meter is investigated. Chapter 7 looks at spread spectrum transmission methods showing how this technique excels in detecting and recovering wanted signals buried in noise.

ACKNOWLEDGEMENTS

I would like to acknowledge certain people who helped me directly and indirectly in producing this book. Many years ago two gentlemen, Tim O'Brien and Paddy Murray introduced me to the world of communications engineering. Much earlier than this though, when I was not quite a teenager, my dad Eddie Tobin (Wizard of the G-banjo), showed me how to build a crystal set and probably this, more than anything else, left me with a passion for this subject which continues to this day. I would like to thank Dr Gerald Farrell, Head of School of Electronics and Communications Engineering DIT, for his help and encouragement. Last, but not least, I would like to thank Joel Claypool of Morgan and Claypool publishing for taking on my five books.

CHAPTER 1

Fourier Analysis, Signals, and Bandwidth

1.1 DIGITAL SIGNALS

Data signals on a limited-bandwidth communications channel are distorted, attenuated, phase shifted, and the ever-present noise added at different stages in the transmission path. There are two basic communication channels: free space, where passband techniques transport the data in an unguided fashion (microwave point-to-point systems are not really in this category), and guided transmission systems such as cables, optical fibers, coax, etc. For example, signals received from an unguided free-space channel are a composite signal comprising signals reflected from buildings, structures, i.e., reflected multipath signals. We should be thankful for this phenomenon because without it we might not receive signals on our mobile phones where buildings, walls, etc. would block line-of-sight to the transmitter.

However, multipath signals also cause signal distortion and have to be removed in the receiver. The channel also distorts the transmitted signal with energy from each transmitted pulse "leaking" into the next transmitted pulse making it difficult for correct pulse identification in the receiver. This is a phenomenon called intersymbol interference (ISI) and the receiver cannot make correct decisions whether a pulse is present or not. One technique for reducing ISI and producing a maximum transmission symbol rate is to shape, or predistort, pulses before transmission. This removes high frequencies from the signal prior to transmission. Fig. 1.1 shows the elements of a digital transmission system.

The encoded data from the modulator is prefiltered before transmission using a root-raised cosine (RCC) filter and attempts to mimic the ideal brick-wall Nyquist filter. At the

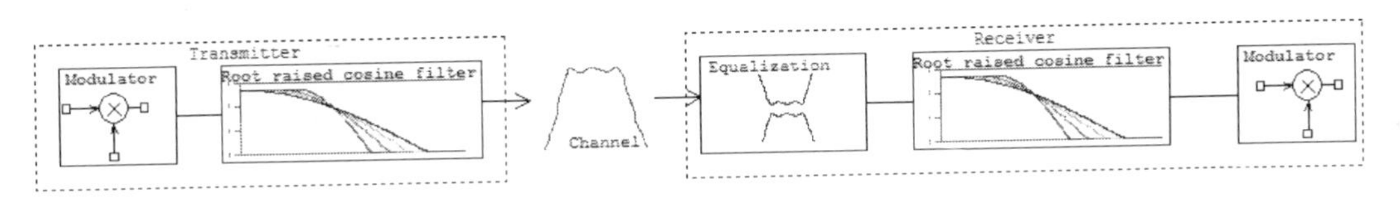

FIGURE 1.1: Transmission system.

receiver is another RCC filter so that the overall path response has a raised cosine filter response. These filters are implemented using digital signal processors that can implement high-order cosine filters that approach the ideal filter response (8). However, transmission path characteristics are generally unknown and vary with time, so adaptive equalizers in the receiver track these changes and subsequently change the filter characteristics as needed. The equalizer flattens the channel response thus ensuring that the RCC filter works efficiently. After equalization, the received signal is root-raised cosine filtered and the demodulator can then make decisions to determine if a pulse is present or not. It does this by sampling the received signal at the center of a pulse and uses logic circuits to ascertain the presence or absence of a pulse.

1.2 BANDWIDTH

In analog communications system, bandwidth is defined as the range of frequencies over which a signal may be transmitted and received with reasonable fidelity, or minimum errors, if the signal is digital. The −3 dB bandwidth is defined as that frequency where the signal is attenuated to half of the value in the passband region. However, this is only one of many bandwidth definitions used in a digital communication channel. For example, data signals have sinc-shaped frequency spectra and one bandwidth definition is the width of the first spectral lobe measured in the spectrum. A transmission line channel has low-pass filtering characteristics and will distort a transmitted pulse by causing it to spread out in time. In the receiver, these smeared pulses will overlap making it difficult in the receiver to ascertain whether a pulse is present or not. We need to examine the shape and spectrum of different types of pulses and see how, by reshaping them using certain filters, we can minimize the problems of symbol interference (ISI). To examine channel limitations, we consider the spectrum of data signals using Fourier series/Fourier transform analysis. The Fourier series is used for periodic functions, whereas the Fourier transform (FT) is for nonperiodic functions. Thus, the FT is a generalization of the Fourier series.

NOTE

1.3 PULSE SPECTRA FOR DIFFERENT PULSE WIDTHS AND PERIOD

Way cool

Signal spectra are examined by selecting the fast Fourier transform (**FFT**) **PROBE** icon after simulation. The objective of this experiment is to investigate the effect on the overall shape of the frequency spectrum when the pulse width and period are changed. Set the **VPULSE** generator part parameters as shown in Fig. 1.2.

Be careful about the rise and fall time parameters as simulation times are increased if these values are too small. The **VPULSE** generator parameters are shown in Fig. 1.3.

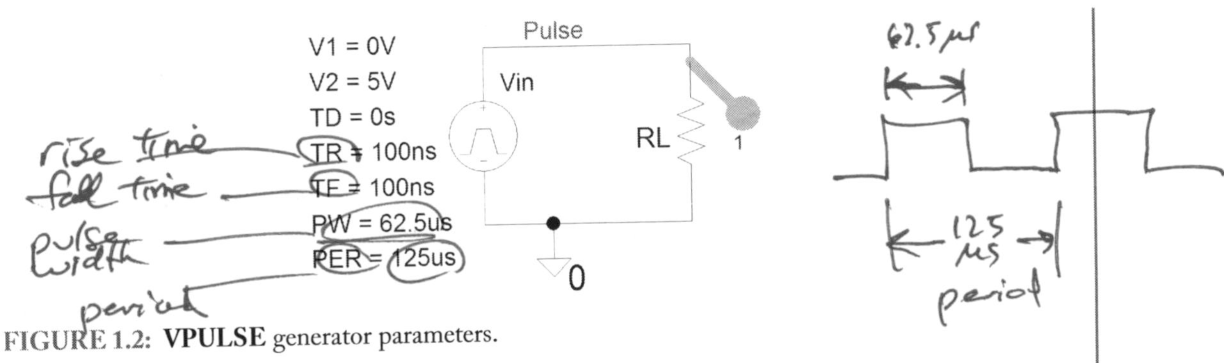

FIGURE 1.2: **VPULSE** generator parameters.

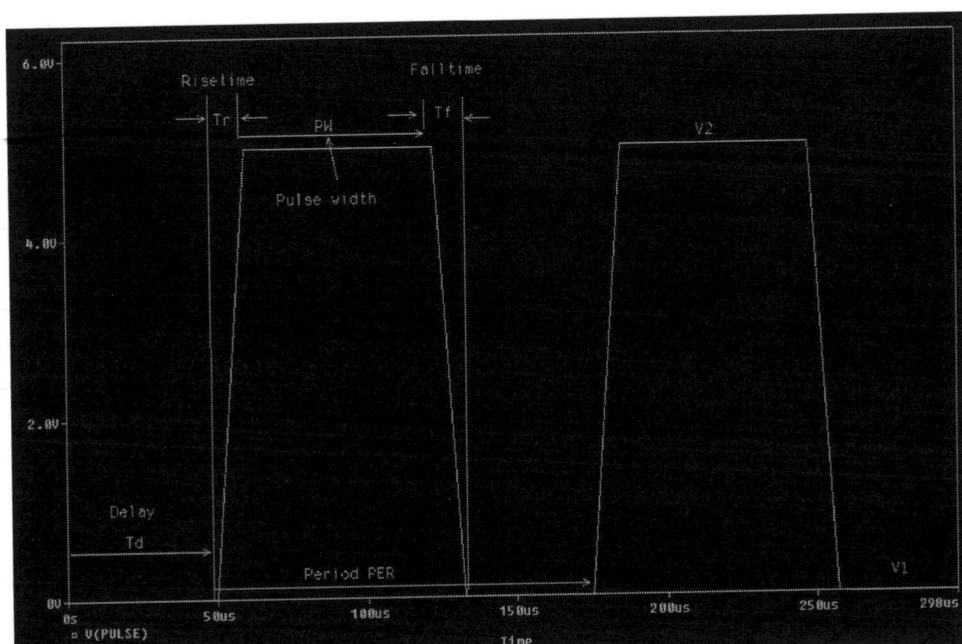

FIGURE 1.3: Pulse generator parameters.

1.3.1 Average and RMS Pulse Power

The average power for a pulse waveform with pulse width τ and period T is connected across a resistance, R,

$$P = \frac{1}{T}\int_0^{\tau} \frac{V^2}{R} dt = \frac{1}{TR}\left[V^2 t\right]_0^{\tau} = \left(\frac{\tau}{T}\right)\frac{V^2}{R}\,\mathrm{W}. \tag{1.1}$$

The RMS value for a pulse with $\tau = 10$ us and period $T = 125$ us is

$$V_{\mathrm{RMS}} = \sqrt{\frac{\tau}{T}}V = \sqrt{\frac{10\,\mathrm{u}}{125\,\mathrm{u}}}5 = 1.414\ \mathrm{V}. \tag{1.2}$$

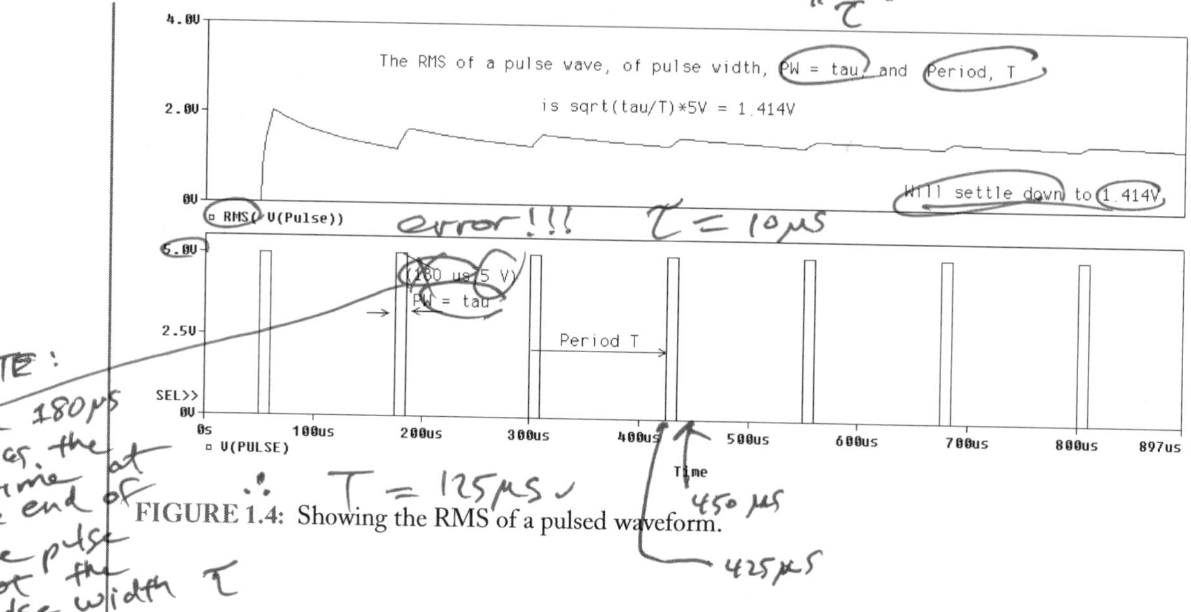

FIGURE 1.4: Showing the RMS of a pulsed waveform.

Thus, the RMS value for a square wave, where $\tau = T/2$, is $V/\sqrt{2}$; the same as a sinusoidal signal. The transient parameters are as follows: **Output File Options/Print values in the output file** = 1ms, **Run to time** = 20ms, **Maximum step size** = 0.01 µs. Press **F11** to simulate and from **PROBE** add another axis using **alt PP** and add the root mean square of the load voltage rms(V(RL:2)) as shown in Fig. 1.4. Note that setting the **Print values in the output file** to a higher value than the default value speeds up the simulation but it must be less than the **Run to time**.

To obtain good resolution in the spectra, we must ensure that the **Run to time** is large and the **Maximum step size** is small. This results in longer simulation times, however. Reset the pulse period to a square wave pulse, i.e., **PER** = **PW**, and simulate. Click the **FFT** icon and measure the spectral components shown in Fig. 1.5.

A rectangular pulse, with amplitude h and width τ has a sinc-shaped spectrum $\sin 2\pi f t / 2\pi f t$ with magnitude $h\tau$ at DC and spectral nulls at frequencies $f = 1/\tau, 2/\tau, 3/\tau$, etc. If the area $h\tau$ of the voltage pulse is constant and the width is decreased, then the spectrum widens but the peak at DC remains at $h\tau$.

1.3.2 Unsynchronizing PROBE–Plot Axis

We may display signals in time and frequency simultaneously using the following methods. The first method is the simplest: from **PROBE**, select the **Windows** menu, create a **New window** and **Tile Vertically**. Copy the time variable across to the new window and click the **FFT** icon. The second method uses the **Unsynchronizing** axis facility in **PROBE/Plot**. After simulation,

FIGURE 1.5: Spectrum of a square wave.

we should observe the square wave signal across the load resistance. Press in sequence, the short-cut keys **alt PP** to produce an extra plot above the original. Position the mouse on the variable v(Rload:1) at the bottom where it should turn red. Apply the short-cut keystrokes **ctrl C** and **ctrl V** to copy and paste the signal into the new plot. To display the two signals in the time and frequency domains simultaneously, as shown in Fig. 1.6, select from the **Plot**

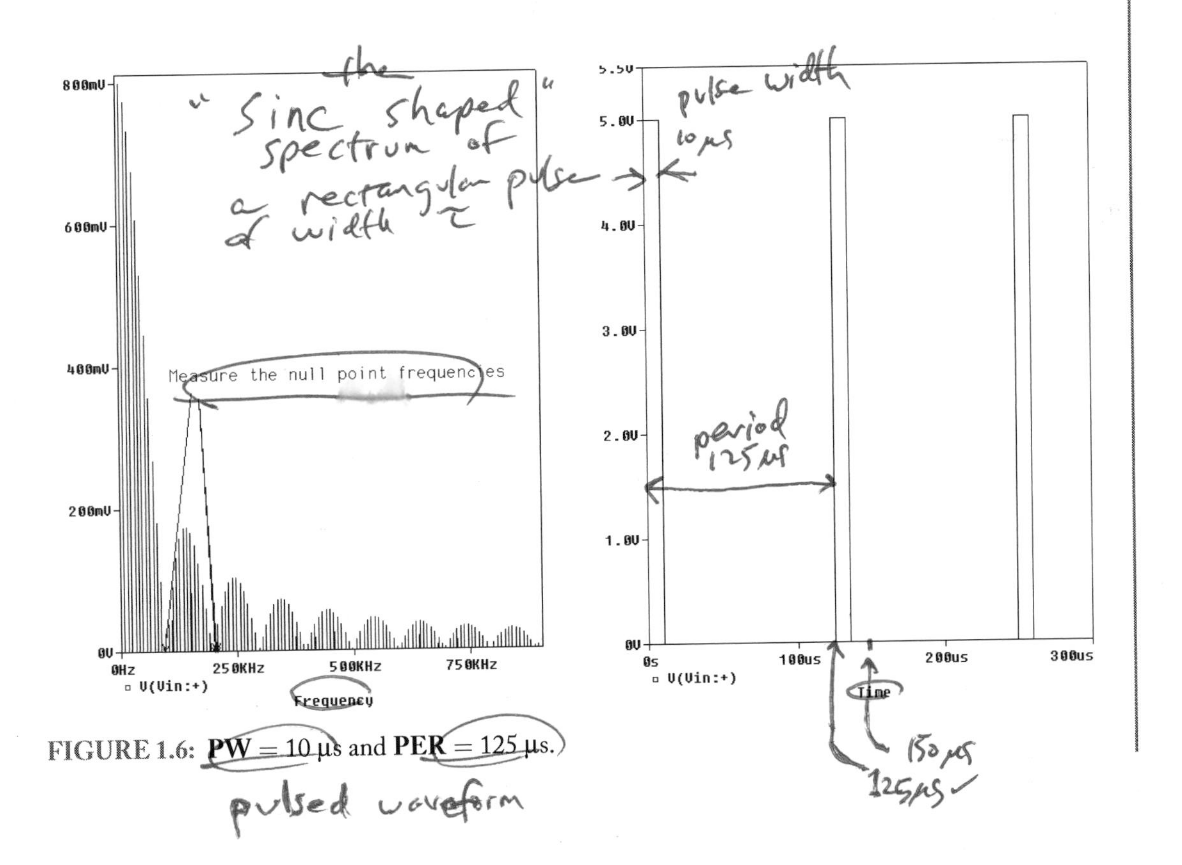

FIGURE 1.6: **PW** = 10 μs and **PER** = 125 μs.

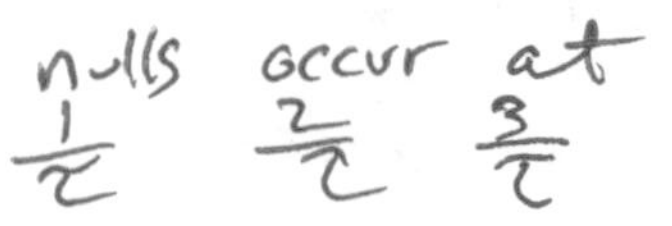

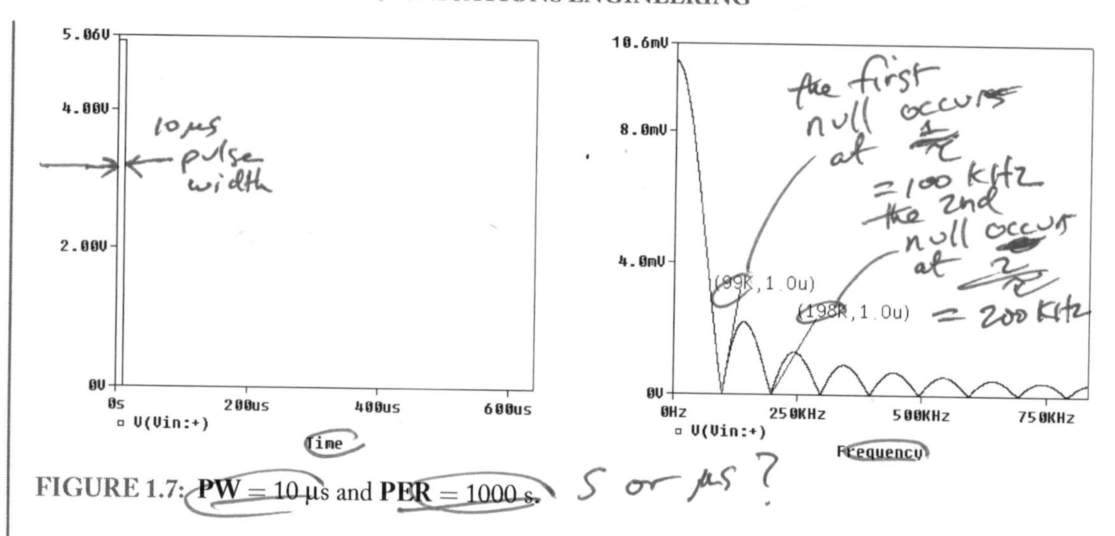

FIGURE 1.7: **PW** = 10 µs and **PER** = 1000 s.

menu, **Unsynchronize Plot** and click the **FFT** icon to display the pulse spectrum (sinc- shaped discrete lines) together with the time-domain pulse train.

The spectrum in Fig. 1.7 shows that the sinc lobes are continuous when an impulse is created by making the **VPULSE** generator period (**PER**) much longer when compared to the pulse width (**PW**). The resultant spectrum flattens out around the origin (DC).

If we could make the impulse infinitely thin then the spectrum would flatten out to infinity and the spectral components will also get smaller. To investigate the effects on the spectrum, reduce the pulse width, increase the pulse period and simulate.

1.3.3 Fourier Transform

The Fourier transform (FT) is a generalization of the Fourier series as outlined in Section 1.3.4 to examine periodic time function, and is used when a signal is not periodic. Applying the FT to a square wave, of duration τ and amplitude A results in a spectrum which is sinc-shaped. Consider the following analysis:

$$v(f) = \int_{-\infty}^{\infty} v(t)e^{-j2\pi ft}dt = \int_{-\tau/2}^{\tau/2} Ae^{-j2\pi ft}dt = A\left[\frac{e^{-j2\pi ft}}{-j2\pi f}\right]_{-\tau/2}^{\tau/2}$$

$$= A\tau\frac{(e^{j\pi f\tau} - e^{-j\pi f\tau})}{j2\pi f\tau} = A\tau\frac{\sin \pi f t\tau}{\pi f t\tau} = A\tau \sin c(\pi f\tau). \quad (1.3)$$

The power spectral density (PSD) is defined as

$$\{v(f)\}^2 = (A\tau)^2\left(\frac{\sin \pi f\tau}{\pi f\tau}\right)^2. \quad (1.4)$$

From (1.4), we see that the PSD has a maximum value of $(A\tau)^2$ at 0 Hz (DC), and the first null (a zero crossing) occurs at $\sin \pi f\tau = 0$ (a frequency $f = 1/\tau$), with 90% of the signal energy in the first lobe of the spectrum. As the pulse narrows, the main spectral lobe widens and increases the channel bandwidth requirements. Thus, transmitting infinitely thin digital pulses with no distortion requires a channel with an infinite bandwidth and a linear phase response—conditions that are not physically realizable. We may now apply the inverse Fourier transform to a narrow pulse processed through an ideal low-pass filter with cut-off frequency $f_c \ll 1/\tau$ to get back to the time domain:

$$v(t) = \int_{-\infty}^{\infty} v(f)e^{j2\pi ft}df = A\tau \int_{-f_c}^{f_c} \frac{\sin \pi ft}{\pi ft} e^{j2\pi ft}df. \tag{1.5}$$

The sinc function $\frac{\sin \pi f\tau}{\pi f\tau}$ is 1 (prove this by applying L'Hopital's rule) for small values of $f_c\tau$, so (1.5) becomes

$$v(t) = A\tau \int_{-f_c}^{f_c} e^{j2\pi ft}df = A\tau \left[\frac{e^{j2\pi ft}}{j2\pi t}\right]_{-f_c}^{f_c} = A\tau \frac{(e^{j2\pi f_c t} - e^{-j2\pi f_c t})}{j2\pi t}$$

$$= 2Af_c\tau \frac{\sin 2\pi f_c t}{2\pi f_c t} = 2Af_c\tau \,\mathrm{sinc} 2\pi f_c t. \tag{1.6}$$

A time-domain sinc signal is simulated using an equation defined in an **ABM** part as shown in Fig. 1.8. Enter the equation 2*A*fc*tau*sin(2*pi*fc*(time-k))/(2*pi*fc*(time-k)) by clicking the pi value to bring you into the **Value** box of EXP1. You could simplify this equation by canceling out the $2f_c$ terms (they were left in to avoid confusion). The first zero occurs at $\sin 2\pi f_c t = 1$, or $t = 1/2f_c$, and other zero crossings at $n/2f_c$. The limited channel bandwidth causes pulses in the transmitted pulse stream to spread in time and overlap so that the receiver might not be able to distinguish between 0 and 1. This is called intersymbol interference (ISI) and together with the presence of noise and other interference signals produces errors in the bit stream. The **PARAM** part is used to define constants including a delay factor, k, which is

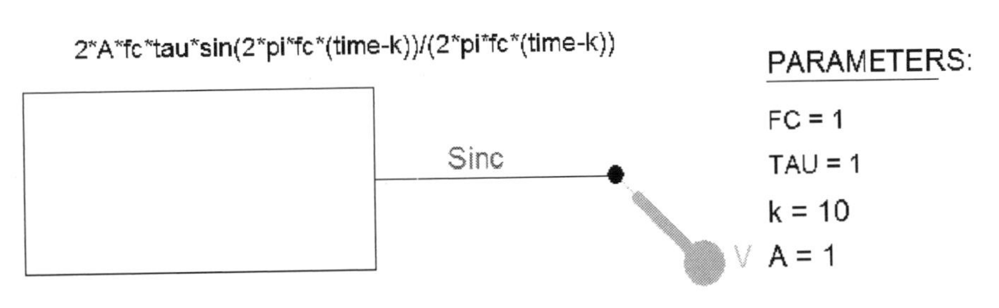

FIGURE 1.8: Sinc pulse.

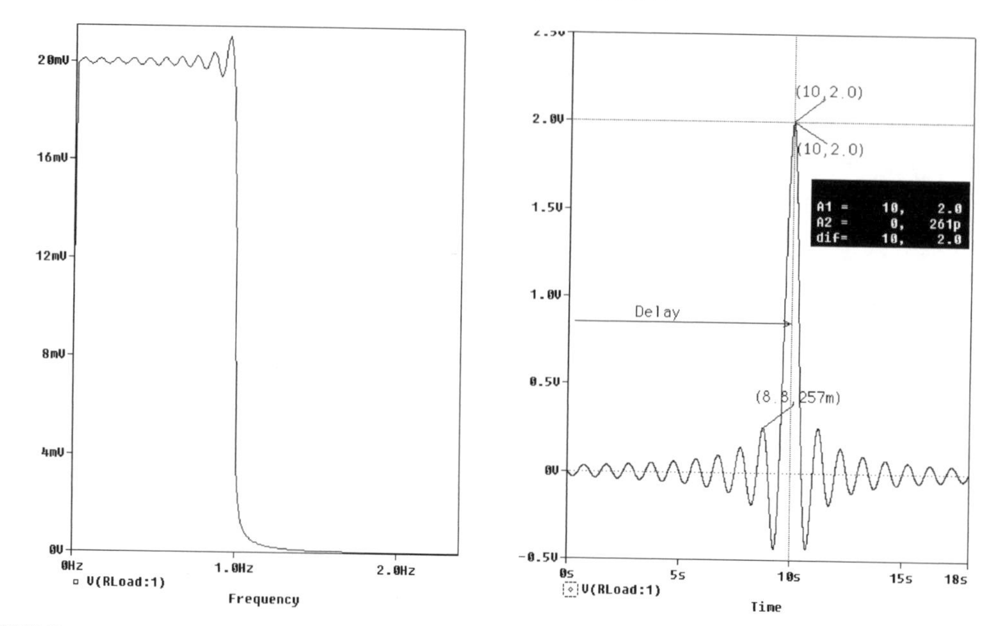

FIGURE 1.9: Sinc signals and pulses.

included to realize a causal sinc function (note that it is not now necessary to define pi as was required in previous editions of PSpice).

Set the transient parameter **Run to time** = 100 s. The sinc-shaped spectrum is observe in Fig. 1.9. Click the **FFT** icon and use the magnifying icon to observe the duality that exists between the sinc waveforms and pulses. We also observe the Gibbs effect occurring in the passband region.

1.3.4 Fourier Series

We can use the Fourier series to synthesize a function, $f(t)$, with period T and fundamental frequency f_0, as the sum of weighted sine and cosine functions and a DC term. For example, a square wave can be reconstructed accurately as more and more cosine and sine components are added. The signal in Fig. 1.5 is defined as

$$x(t) = \begin{cases} A \rightarrow -\tau/2 < t < \tau/2 \\ 0 \rightarrow \tau/2 < t < (T - \tau/2). \end{cases} \tag{1.7}$$

The Fourier series for a periodic function, $f(t)$, is

$$f(t) = A_0 + \sum_{n=1}^{N} (a_n \cos 2\pi n f_0 t + b_n \sin 2\pi n f_0 t). \tag{1.8}$$

Here A_0 is a DC term and f_0 is the fundamental frequency equal to $1/T$. The DC coefficient is calculated for the signal defined in (1.7) as

$$A_0 = \frac{1}{T}\int_0^T f(t)dt = \frac{1}{T}\int_{-\tau/2}^{\tau/2} Adt = \left[\frac{A}{T}t\right]_{-\tau/2}^{\tau/2} = \frac{A\tau}{T}. \tag{1.9}$$

If the signal is a square wave, then $T = 2\tau$, so that (1.9) is $A_0 = A/2$. The a_n coefficient is determined as

$$a_n = \frac{2}{T}\int_0^T f(t)\cos 2\pi n f_0 t dt \rightarrow a_n = \frac{2}{T}\int_{-\tau/2}^{\tau/2} A\cos 2\pi n f_0 t dt = \frac{2A}{2\pi n f_0 T}\left[\sin 2\pi n f_0 t\right]_{-\tau/2}^{\tau/2}. \tag{1.10}$$

Since $\sin(-x) = -\sin(x)$ and $f_0 = 1/T$,

$$\begin{aligned} a_n &= \frac{2A}{2\pi n f_0 T}\left[\sin(2\pi n f_0 \tau/2) - \sin(-2\pi n f_0 \tau/2)\right] \\ &= \frac{2A}{\pi n f_0 T}\left[\sin(2\pi n f_0 \tau/2)\right] = \frac{2A\tau}{T}\frac{\sin(\pi n\tau/T)}{\pi n\tau/T}. \end{aligned} \tag{1.11}$$

The b_n coefficient is calculated as

$$b_n = \frac{2}{T}\int_0^T f(t)\sin 2\pi n f_0 t dt = \frac{2}{T}\int_{-\tau/2}^{\tau/2} A\sin(2\pi n f_0 t)dt = \frac{-A}{\pi n f_0 T}\left[\cos(2\pi n f_0 t)\right]_{-\tau/2}^{\tau/2} \tag{1.12}$$

$$b_n = \frac{-A}{\pi n}\left\{\cos(2\pi n f_0 \tau/2) - \cos(-2\pi n f_0 \tau/2)\right\} = 0, \tag{1.13}$$

since $\cos(x) = \cos(-x)$. Thus the Fourier series expansion is written as

$$f(t) = \frac{A\tau}{T} + \frac{2A\tau}{T}\sum_{n=1}^{\infty}\frac{\sin(\pi n\tau/T)}{\pi n\tau/T}\cos(2\pi n f_0 t). \tag{1.14}$$

The schematic in Fig. 1.2 has the following parameters: **VPULSE** generator pulse width **PW** = 1 μs equal to half the pulse period **PER** = 2 μs, thus the fundamental frequency of the pulse signal is 500 kHz. To display harmonic information in the output file, select the simulation menu and tick **Perform Fourier Analysis, Center Frequency** = 500 kHz, **Number of Harmonics** = 10, and **Output Variables:** = v(squarewave). Select **Analysis/Examine Output** to display the harmonic analysis information at the end of the ".out" text file. Press **F11** and examine from **PROBE/View** the harmonic details such as amplitude, frequency, and phase in the output file as shown in Fig. 1.10. Write down the frequency, amplitude, and phase for each of the odd harmonics, i.e., 1, 3, 5, 7, etc.

FOURIER COMPONENTS OF TRANSIENT RESPONSE V(SQUAREWAVE)

DC COMPONENT = 2.505000E+00

HARMONIC NO	FREQUENCY (HZ)	FOURIER COMPONENT	NORMALIZED COMPONENT	PHASE (DEG)	NORMALIZED PHASE (DEG)
1	5.000E+05	3.183E+00	1.000E+00	-1.620E+00	0.000E+00
2	1.000E+06	1.000E-02	3.141E-03	9.000E+01	9.324E+01
3	1.500E+06	1.062E+00	3.335E-01	-4.863E+00	-2.428E-03
4	2.000E+06	1.000E-02	3.141E-03	9.000E+01	9.648E+01
5	2.500E+06	6.378E-01	2.003E-01	-8.113E+00	-1.212E-02
6	3.000E+06	1.000E-02	3.141E-03	9.000E+01	9.972E+01
7	3.500E+06	4.563E-01	1.434E-01	-1.137E+01	-3.388E-02
8	4.000E+06	1.000E-02	3.141E-03	9.000E+01	1.030E+02
9	4.500E+06	3.558E-01	1.118E-01	-1.465E+01	-7.239E-02

TOTAL HARMONIC DISTORTION = 4.294822E+01 PERCENT

FIGURE 1.10: End of the output file.

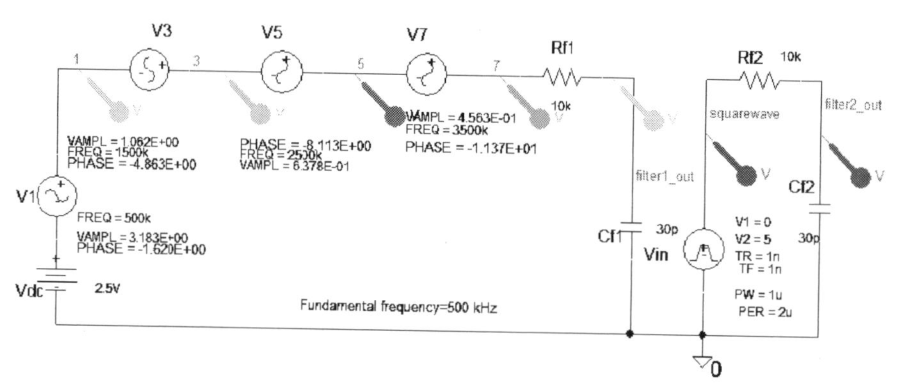

FIGURE 1.11: Synthesizing a square wave.

We will now attempt to reconstruct this square wave using the fundamental frequency and the first three odd harmonics shown in the output file. In Fig. 1.11, connect the generators in series and set the frequency and amplitude values for each VSIN generator part as shown. Set the **Analysis** tab to **Analysis Type: Time Domain (Transient), Run to time** = 0.5ms, **Maximum step size** = 1 μs, and press **F11** to simulate. The **Run to time** is set to a much larger value than the square wave period to achieve good **FFT** resolution. After simulation we can see how a square wave is synthesized by adding the odd harmonics. The sum of these generators is connected to a low-pass CR filter to mimic the limited low-pass filter bandwidth characteristics of a transmission line channel. Adding more generators of course would produce a better approximation to the Vin square wave included in the schematic. The synthesized and generator square wave signals are shown in Fig. 1.12.

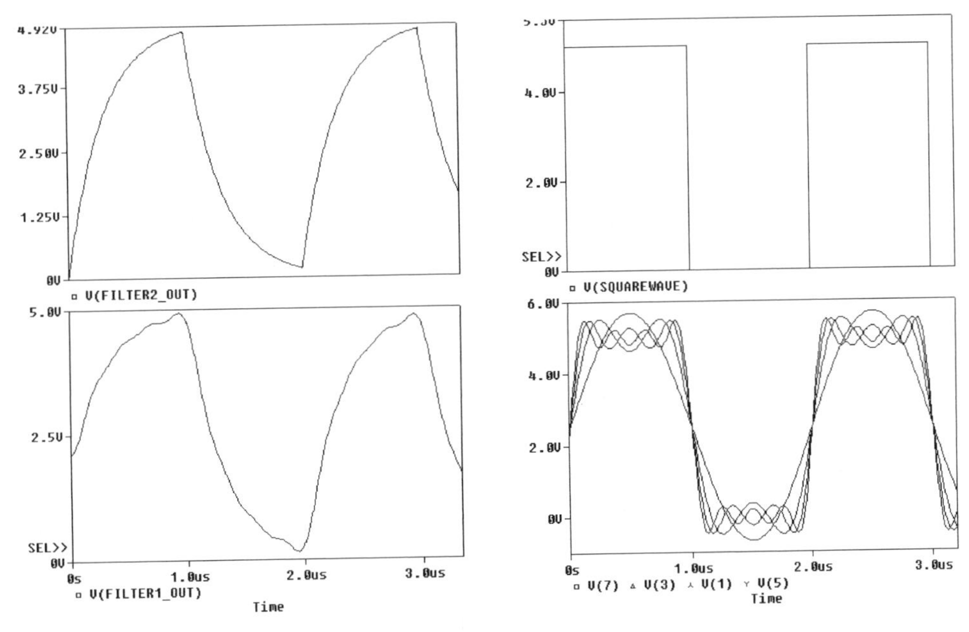

FIGURE 1.12: Approximation to the square wave.

Select the **Net Alias** icon and enter **OUT** in the name box. Drag the little box to the junction of the resistor–capacitor wire segment and place it on the wire.

1.4 THE VECTOR PART

A **VECTOR** part records digital signals at a location when placed on a schematic. Placing a **VECTOR** part on a schematic records digital signals only and creates a file "myfilename.vec" that may be applied as an input signal source using the **FileStim** generator part. This is useful for breaking larger digital circuits into smaller subcircuits to satisfy the limitations of the evaluation version. PSpice creates two columns of time–voltage pairs in the **VECTOR** file. To see how this part is used, create a pseudorandom binary sequence (PRBS) generator using the circuit shown and place the **VECTOR** part (a little square box symbol) as shown in Fig. 1.13.

Fig. 1.14 shows the necessary parameters to be set when simulating digital circuits. The flip-flops must be initialized into a certain state. Select the simulation setting menu and then select the **Options** tab. In the **Category** box select **Gate-Level Simulation**. This will then show the **Timing Mode**- select **Typical**. Make sure to select **Initialization all flip-flops to 1**. Select and right click the **VECTOR** part. Select **Edit Properties** and the directory and file name are specified in the **FILE** parameter. Select **FILE** and in the second column enter C:\Pspice\Circuits\signalsources\data\prbs1.txt.

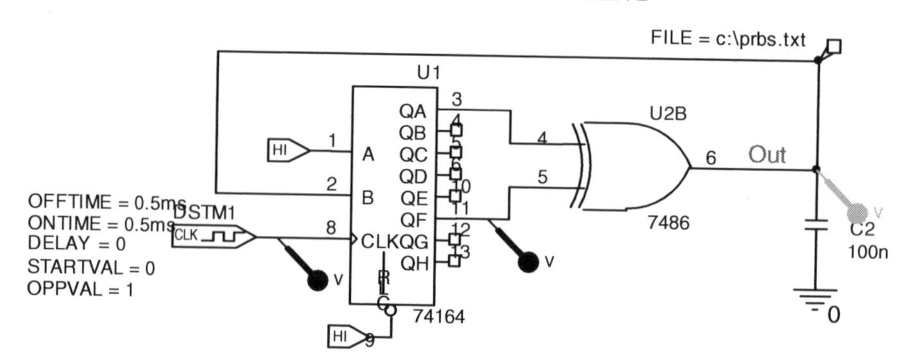

FIGURE 1.13: Use of **VECTOR** part.

FIGURE 1.14: Initialize all flip-flops.

Other **VECTOR** parameters are as follows:

- POS: this is the column position in the file with values ranging from 1 to 255.
- FILE: the location and file name must be specified, e.g., C:\signalsources\prbs.txt.
- RADIX: valid values for **VECTOR** symbol attached to a bus are B[inary], O[ctal], and H[ex].
- BIT: if the **VECTOR** symbol is attached to a wire, the bit position within a single hex or octal digit.
- SIGNAMES: this is the wire segment name in the file header where the **VECTOR** part is connected.

The PRBS clock and output signals are shown in Fig. 1.16.

prbs.txt - Notepad

File Edit Format View Help

```
* Created by PSpice
out

0s              0
1.000036ms      1
2.000029ms      0
3.000036ms      1
4.000029ms      0
5.000036ms      1
7.000029ms      0
9.000028ms      1
```

FIGURE 1.15: The file created by the **VECTOR** part.

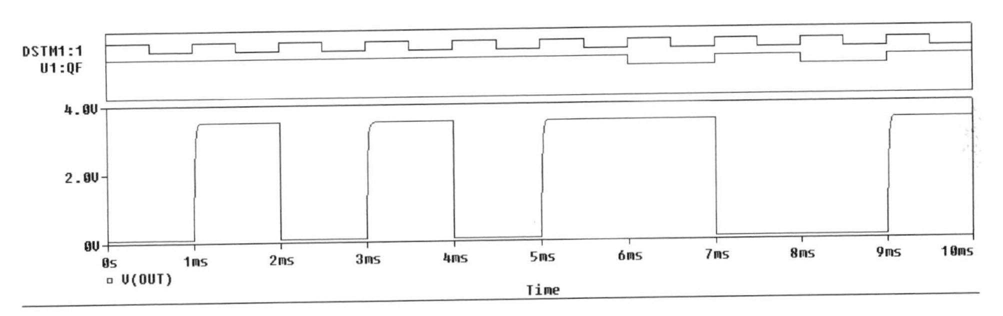

FIGURE 1.16: PRBS output.

The **VECTOR** data file created, as shown in Fig. 1.15, is 01010101 and consists of three parts: (1) * Created by PSpice (a comment line), (2) the header (the wire segment name is called out), and (3) a column pair consisting of time and data level amplitudes (i.e., 1 or 0).

The length of the data file, created using the **VECTOR** part, depends on the transient **Run to time** value.

1.5 EXERCISE

1. Investigate Fourier analysis for Triangle, Sawtooth, and rectified sinusoid signals.

CHAPTER 2
Baseband Transmission Techniques

2.1 BASEBAND SIGNALS

Fig. 2.1 shows a digital data sequence $x(n)$ encoded onto a line code $s(t)$ for transmission over a limited-bandwidth channel. Signals not carrier modulated are referred to as baseband signals. The encoded data, in unipolar or bipolar format, is sent directly to the channel so that the data spectrum starts at DC.

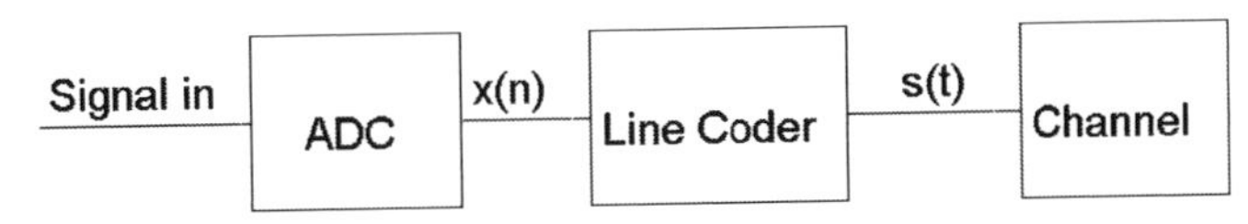

FIGURE 2.1: Baseband coder.

Line coding a data stream enables a receiver to extract a timing clock signal from the received signal so that the transmitter and receiver operate in synchronism. The code should produce a signal with a suitable spectrum consistent with minimum bandwidth. For example, long sequences of 0's or 1's in the transmitted sequence produce DC in the spectrum that should be avoided as the public telephone network contains transformer and capacitive-coupled networks that will not transmit DC. Examples of baseband codes are nonreturn to zero (NRZ—also called NRZ-L), Manchester (or biphase, Biϕ), differential Manchester, and alternate mark inversion (AMI). A communication channel has many forms: free-space, twin-pair cable, coax cable, optical fiber, and the bandwidth for each channel is different placing limits on the amount of information it can carry at any one time. Channel noise also limits the information transmission rate and causes errors to be detected in the receiver.

2.2 BASEBAND SIGNAL FORMATS

Baseband signals have two formats: polar and bipolar. The bipolar format has advantages over the polar format because it has no DC content when equal numbers of 0's and 1's occur in the transmitted message signal. The polar format (also referred to as unipolar) contains DC and cannot be transmitted over a telephone network that uses transformer/capacitor coupling. Errors occur in noisy systems where strings of 1's or 0's change the decision threshold making it difficult for the receiver to detect whether a 1, or 0, is present. The bipolar format, for

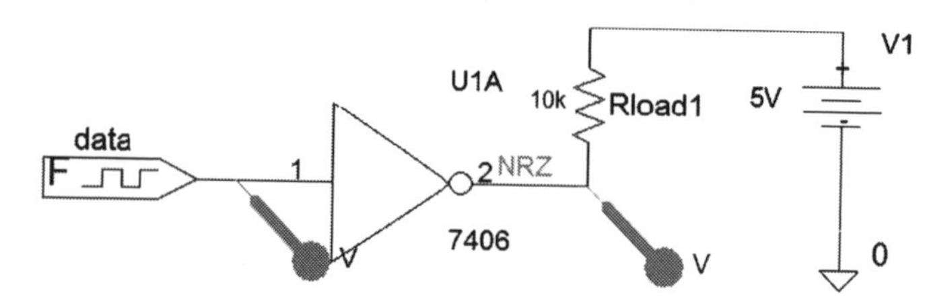

FIGURE 2.2: NRZ production.

the same signal to noise ratio, requires half the average power compared to polar signals. TTL (transistor–transistor logic—same format as nonreturn to zero) represents a logic level 0 (a space) as 0–0.8 V, and a logic level 1 (a mark) as 2–5 V, with output current less than 15 mA.

2.2.1 NonReturn to Zero (NRZ) Coding

The circuit in Fig. 2.2 produces a two-level polar nonreturn to zero (NRZ) data that has poor coding properties and with poor clock extraction properties. Here the pulse width is equal to the bit interval, and we will see from the spectrum plotted in **PROBE** using the **FFT** icon that there is a DC component making it unsuitable for transmitting data over the public switched telephone network (PSTN).

The input data is an ASCII file 00100101...NRZ1.txt created in Notepad© (see Fig. 1.15 in Chapter 1). In this example, the header name is the same as the input wire segment name NRZdata and the header filename must be separated from the first file pair "0s 0" by a blank line.

2.2.2 FileStim Generator

The **FileStim** generator part applies the signal recorded by the **VECTOR** part, examined in Chapter 1, as a digital input source. The **FileStim** generator has two attributes: the first is the **FileName** attribute, where the file location and name is entered, e.g., C:\Pspice\Circuits\signalsources\data\NRZ1.txt; the second attribute is the **SigName** attribute that specifies the name of the wire where it is attached (select a wire and enter a name using the **Net Alias** icon on the right toolbar). If you use **FileStim** with a nondigital signal, you will get "Circuit Too Large" error message displayed. Set the transient **Run to time** to 20ms and simulate to produce the signals shown in Fig. 2.3.

To investigate the NRZ signal spectrum, we need to use a much longer signal, so replace the **FileStim** generator with a 1ms **DigClock**. Set **Output File Options/Print values in the**

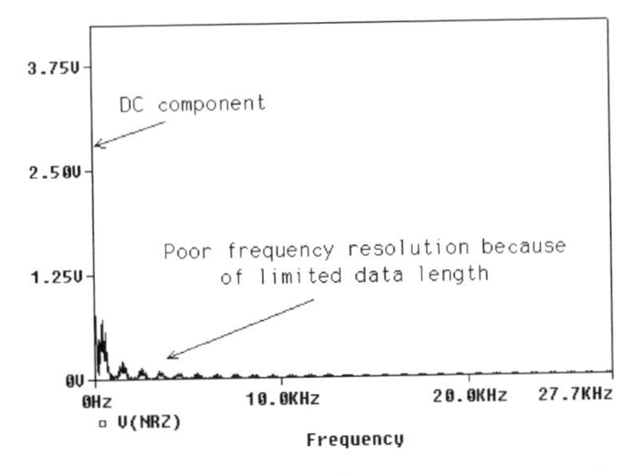

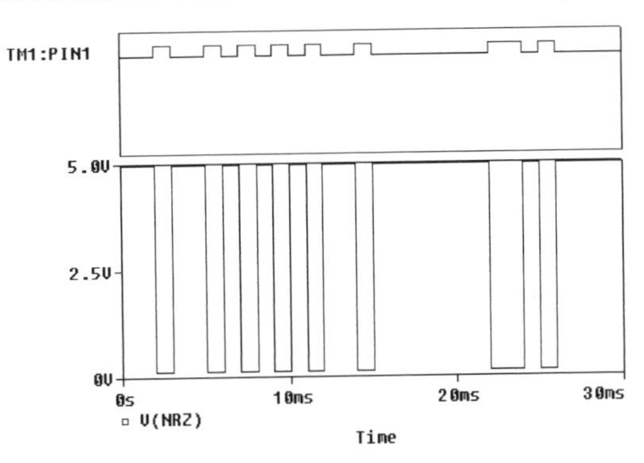

FIGURE 2.3: NRZ signal.

output file to 100ms, **Run to time** to 1 s, and **Maximum step size** = 1 μs and press the **F11** key to simulate. Select the **FFT** icon and observe a 2.5 V DC component present in the spectrum (2.5 V is the average value for a 0–5 V pulse).

2.2.3 NRZ-B

Turning NRZ into bipolar NRZ-B (B stands for bipolar) form eliminates DC from the spectrum. Fig. 2.4 shows how to do this conversion with the input data applied using a **STIM1** part. Set **Run to time** to 1ms, **Maximum step size** (left blank), and simulate with the **F11** key.

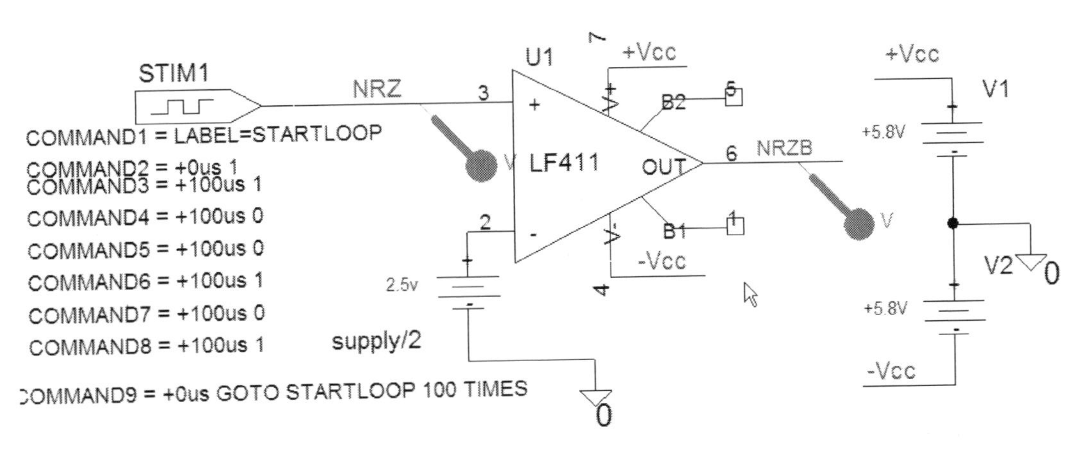

FIGURE 2.4: NRZ bipolar.

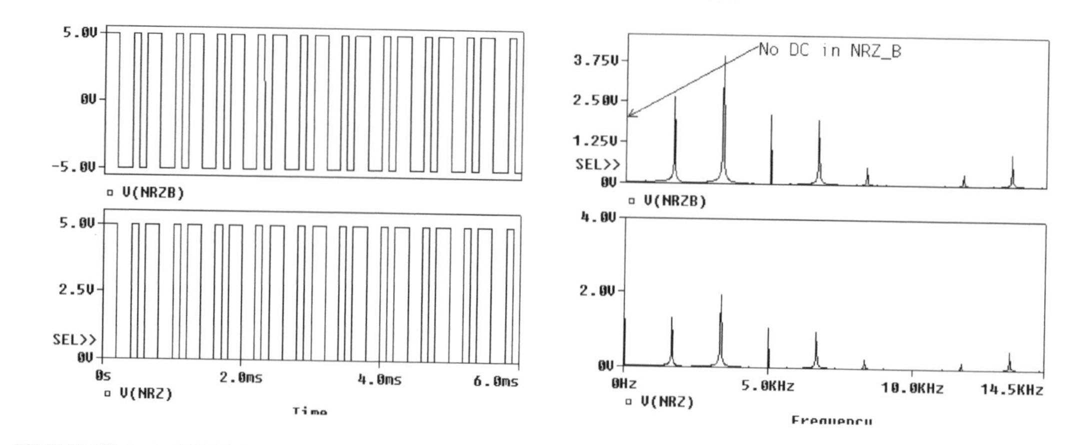

FIGURE 2.5: NRZ-B and NRZ.

Fig. 2.5 is a plot of the NRZ and NRZ-B signals.

From the **PROBE** screen, select the **FFT** icon to see the differences between NRZ-B and NRZ signal spectra.

2.3 RZ ENCODING AND DECODING

Clock signals are generated in the receiver from positive or negative transition levels to zero. The pulses are shifted back by a quarter of a clock period to ensure that the sampling points occur in the centers of the first halves of the bit intervals. This coding requires extra bandwidth because the actual pulse is half the size of the bit interval, and wastes *power* in transmitting a three-level signal. A continuous stream of 1's produces a DC level and causes problems in communication networks that cannot transmit DC. Fig. 2.6 applies the NRZ signal from the

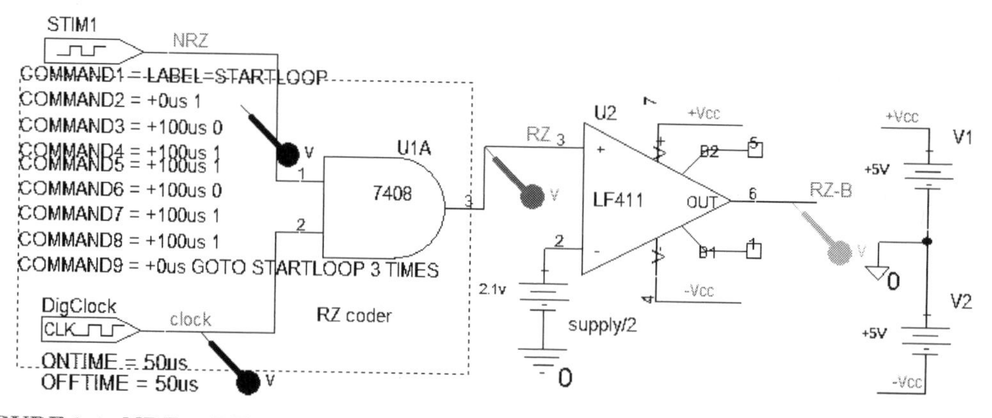

FIGURE 2.6: NRZ to RZ encoder.

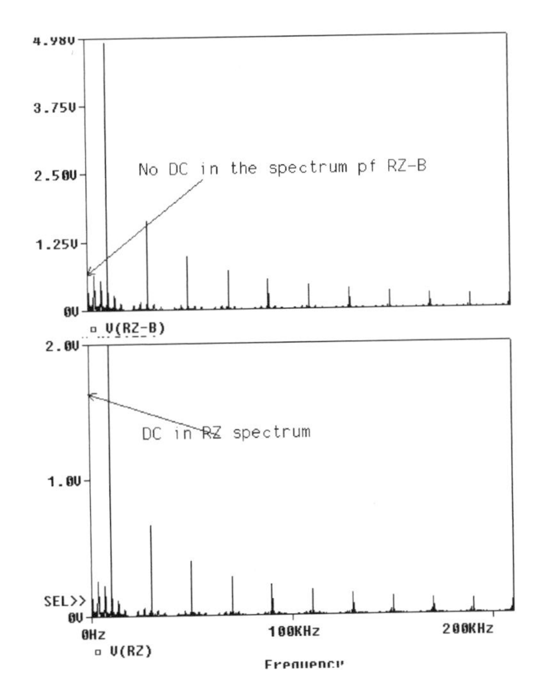

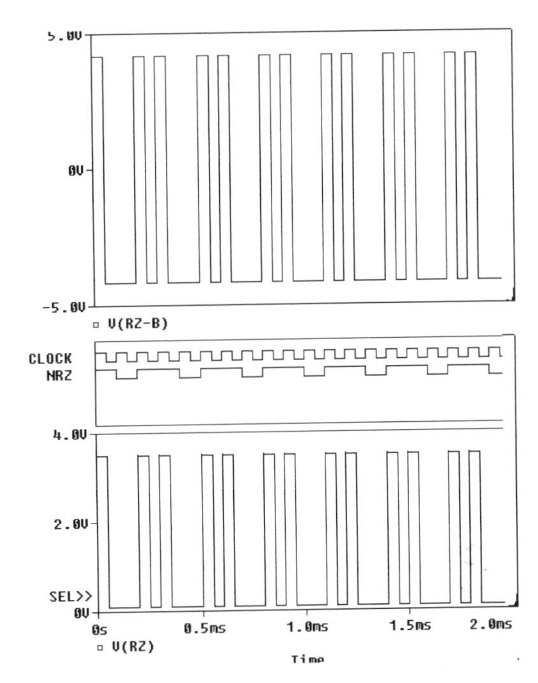

FIGURE 2.7: RZ and RZ-B signals.

previous schematic to produce a Return to Zero (RZ) line coding with a transition in the middle of every bit.

Set **Run to time** = 10ms and simulate with **F11** key. The RZ and RZ-B signals are shown in Fig. 2.7. However, to display the **FFT** of a digital signal in **PROBE** means attaching a resistor from the required node to ground; otherwise you get the message informing you that the **FFT** of a digital trace will not be displayed. The output from the comparator is RZ-B with no DC in the spectrum.

2.3.1 RZ to NRZ Decoder

Fig. 2.8 shows an RZ to NRZ signal decoder. The monostable (74121) pulse width is set by $C1$ and $R1$ connected to pins 10 and 11. However, these components are not modeled in PSpice and are included only for completeness, so we select the 74121 component and set the pulse width to 25 μs.

When using digital devices such as registers, flip-flops, etc., it is important to reset **Initialize all flip flops** from X to 0, or 1. Failing to do this may result in red lines displayed in **PROBE** because the output impedance levels are indeterminate. Select the **Simulation Setting** menu and then select the **Options** tab. In the **Category** box select **Gate-Level Simulation**.

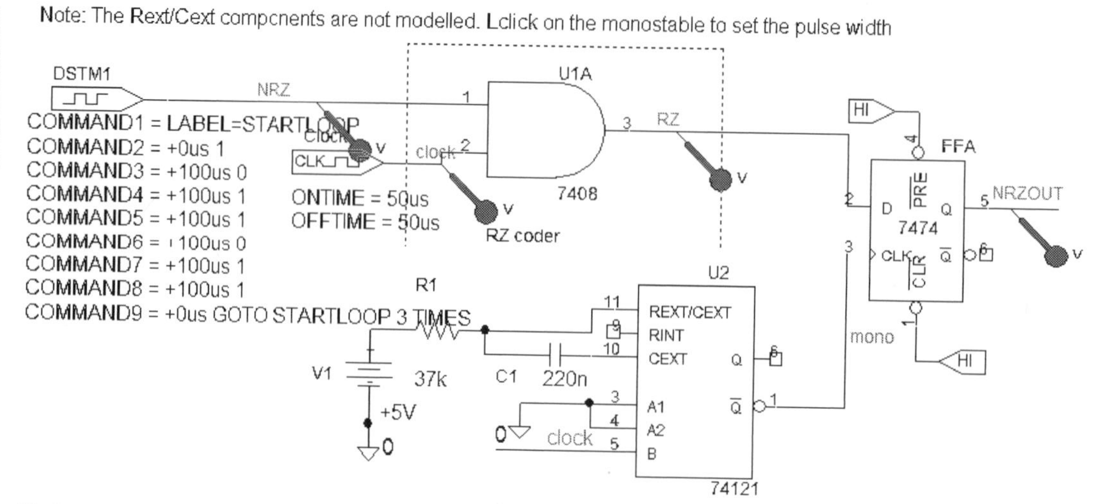

FIGURE 2.8: RZ to NRZ decoder.

This will then show the **Timing Mode** - select **Typical.** Make sure to select **Initialization all flip-flops to 1**. For most simulations, set the **Timing Mode** to **Typical** and the **Default A/D Interface** to **Level 2**. An alternative is to attach an initialization circuit to the CLR pin. Such a circuit may be a *CR* low-pass filter attached to a 5 V DC source that charges up the capacitor on switch-on to the 5 V source. The capacitor will charge up in approximately five time constants to 5 V and is equivalent to applying a HI condition to the IC CLR pin. Set **Output File Options/Print values in the output file** = 100 ns, **Run to time** = 1ms, **Start saving data after** = 0.2ms, and **Maximum step size** = 1 μs, and simulate with the **F11** key. Position the cursor accurately on a leading, or lagging edge, using the icon that moves the cursor to the next digital transition. The NRZ output signal is clearly seen in Fig. 2.9.

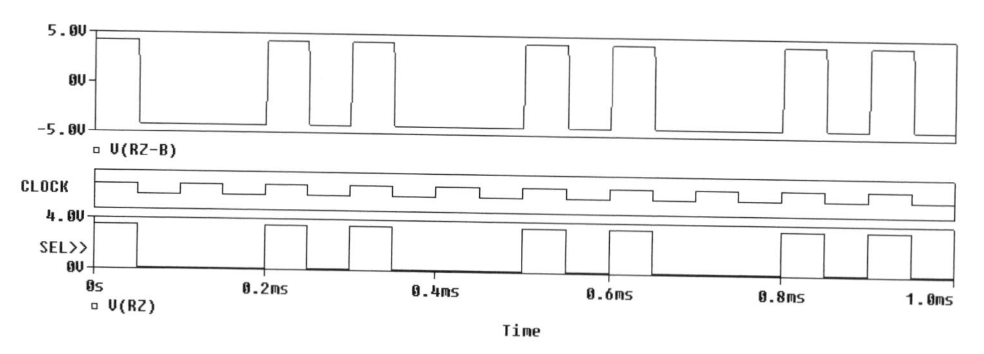

FIGURE 2.9: NRZ output signal.

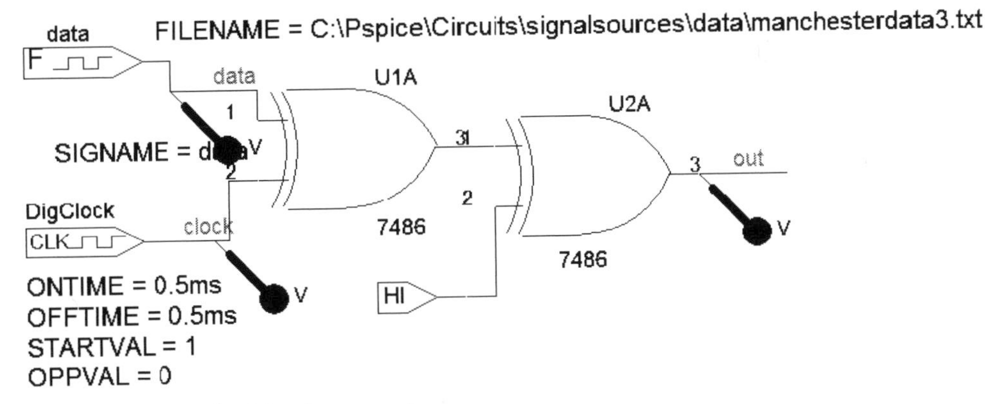

FIGURE 2.10: Unipolar Manchester coder.

2.4 MANCHESTER ENCODING AND DECODING

It is not desirable to have DC content in signals used in the public switched telephone network (PSTN), especially where lines are transformer coupled. The unipolar Manchester code is a two-level code with transitions at the bit centers between two levels (high–low and low–high). Each "1" has a transition from high to low and each "0" has a transition from low to high. The transitions make for easier clock extraction at the receiver. The Manchester code (biphase) requires twice the bandwidth compared to NRZ and NRZ-B and has a DC component. Fig. 2.10 shows a schematic for producing a unipolar Manchester signal.

The file "Manchesterdata3.txt," a data file — "101011100," was created in a text editor Notepad (see Section 1.4) consisting of a header line "* Created by PSpice"; the line segment name data is separated by a space from the time–voltage data. The Manchester data input is applied using a **FileStim** part where the second line, **SigName** = data, is the wire segment name connecting the **FileStim**. **Filename** = C\signalsources\manchesterdata3.txt. Set the **Output File Options/Print values in the output file** to 1ms, **Run to time** to 10 ms, and **Maximum step size** (left blank). Press **F11** to display the signals as in Fig. 2.11.

2.4.1 Manchester Unipolar to BipolarEncoding

Fig. 2.12 shows how to transform a unipolar Manchester code into bipolar form. The bipolar Manchester code has no DC in the spectrum and could be used in the public switched telephone network (PSTN).

Set the **Output File Options/Print values in the output file** to 1ms, **Run to time** to 200ms, and **Maximum step size** (left blank). Simulate with **F11** key to display the bipolar Manchester signals shown in Fig. 2.13.

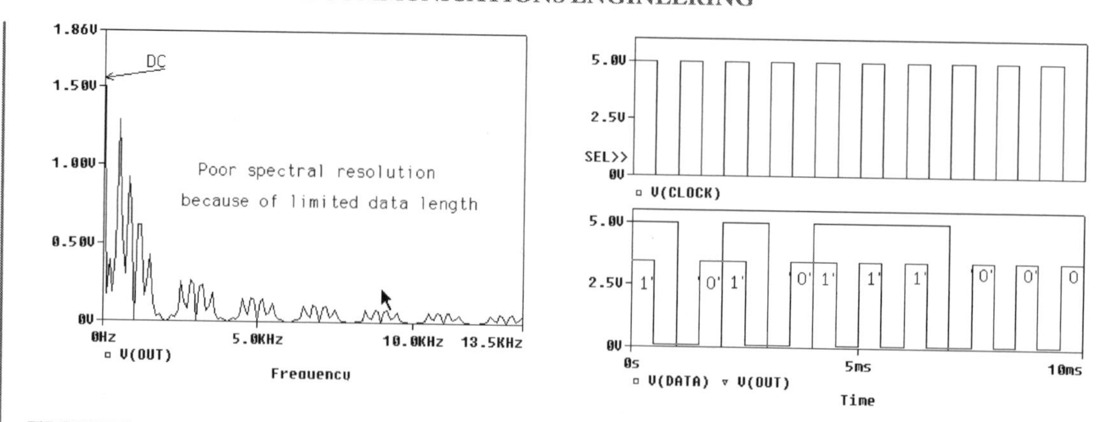

FIGURE 2.11: Unipolar Manchester signals and the DC component.

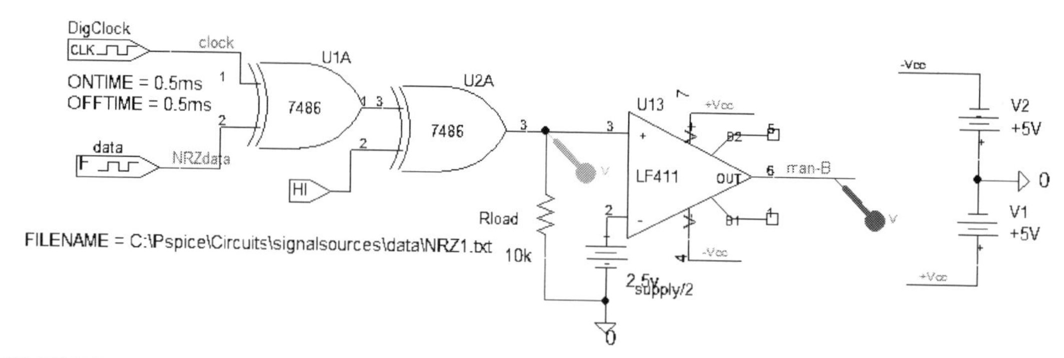

FIGURE 2.12: Bipolar Manchester code.

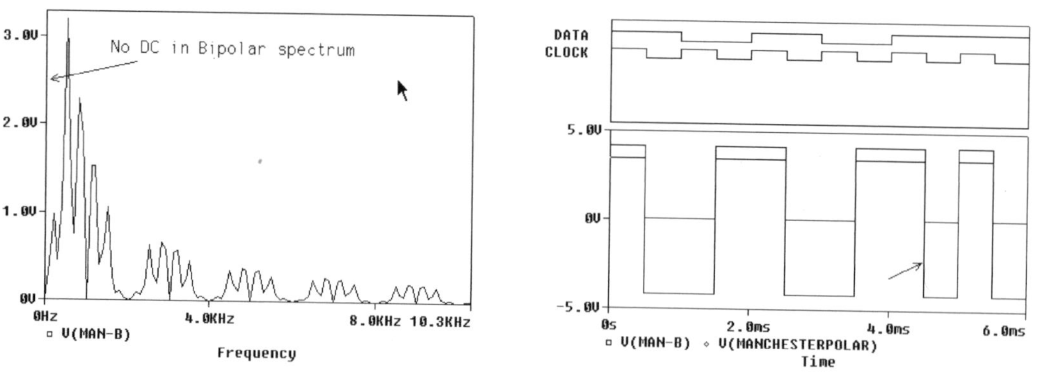

FIGURE 2.13: Bipolar Manchester signals.

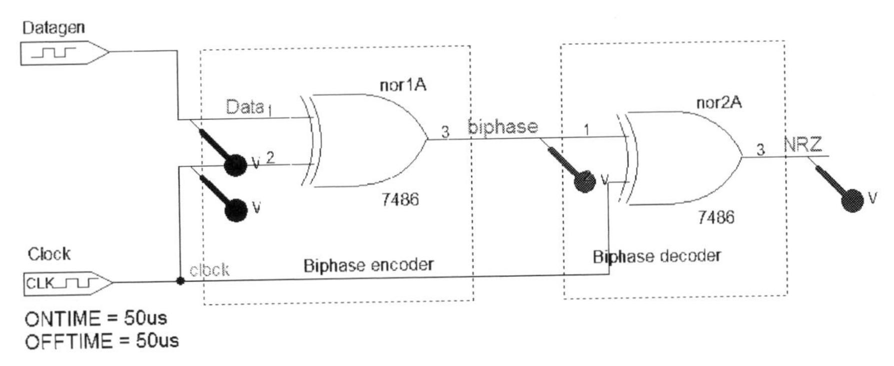

FIGURE 2.14: Biphase production.

An impulse at 3ms occurs because the two input signals are separated slightly in time and when applied to the XOR gate cause an ambiguity resulting in an impulse. Use the magnifying tool to measure the difference between the two pulses (approximately 31 ns between the two edges). Observe how the bipolar Manchester code signal has no DC in the spectrum.

2.4.2 Manchester (Biphase) Decoding

Two XOR gates connected as shown in Fig. 2.14 will generate a biphase signal for the input data applied using a stimulus **STIM1** part named **Datagen**.

Fig. 2.15 shows how looping functions create a repeating input data pattern. The first command is STARTLOOP and ended by GOTO STARTLOOP 100 times. Note that the command GOTO LOOP 2-1 TIMES produces an infinite loop. Enter the test signal "1011011." In this example, command2 is +0us 0, command3 is +100 us 1, etc., and repeated 100 times.

The clock signal is a **DigClock** part with parameters as shown in the schematic. The input line to the NOR device is labeled using the **Net Alias** icon, typing in a name such as **Data** and dragging the little box to the wire segment. This is useful for identifying signals when plotted in **PROBE**. Draw a box around a component, or a group of components, by selecting the box icon from the right-hand menu. After drawing the box, click on a box line to change the line properties if you so wish, i.e., to dotted format as shown above. To observe the DC content in the spectrum, terminate the output with a resistance in order to use the **FFT** function on digital level signals. Set **Output File Options/Print values in the output file** to

COMMAND1	COMMAND2	COMMAND3	COMMAND4	COMMAND5	COMMAND6	COMMAND7	COMMAND8	COMMAND9
LABEL=STARTLOOP	+0us 1	+100us 0	+100us 1	+100us 1	+100us 0	+100us 0	+100us 1	+0us GOTO STARTLOOP 3 TIMES

FIGURE 2.15: STIM1 parameters.

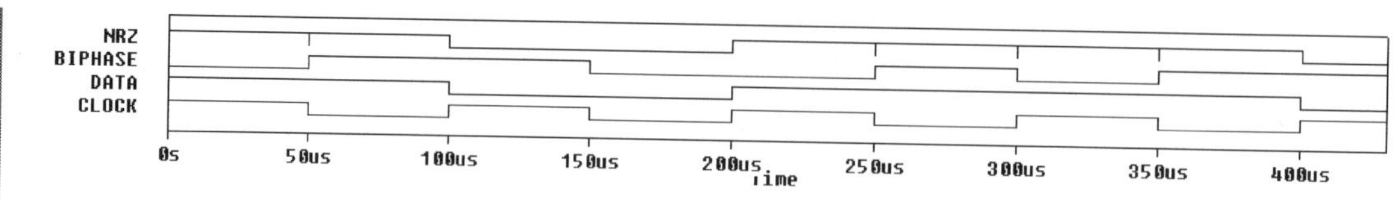

FIGURE 2.16: NRZ signal.

0.1ms, **Run to time** to 1ms, and **Maximum step size** (left blank). Simulate with **F11** key to display the NRZ signals shown in Fig. 2.16.

2.4.3 Differential Manchester Coding

The schematic in Fig. 2.17 uses the data text signal from the previous circuit. Differential Manchester code, used in the Ethernet, has transitions in the middle of the pulse.

Digital warnings after simulation are often associated with **PROBE** screen waveforms that have indeterminate states. The flip-flops must be initialized into a certain state. Select the **Simulation Setting** menu and then select the **Options** tab. In the **Category** box select **Gate-Level Simulation**. This will then show the **Timing Mode** - select **Typical**. In the **Timing Mode** change **Digital Setup** from **Typical** to **Maximum**. The default **Initialize all flip flops** is set to X. Failing to change this will result in the flip-flop being in an undetermined state and shows up in the plot as two red lines, so change **Initialize all flip flops** to 0. The differential Manchester signal in Fig. 2.18 shows transitions occurring in the middle of the pulse that are determined by 1 or 0 and the preceding bit. The transition stays the same as the preceding one, if the current bit is 0, but switches when the bit is 1. Once the start bit is known, all the following bits are obtained, which gives this code an advantage. For example, this code ensures no problems if wires are mistakenly reversed in a connector at the time of installation.

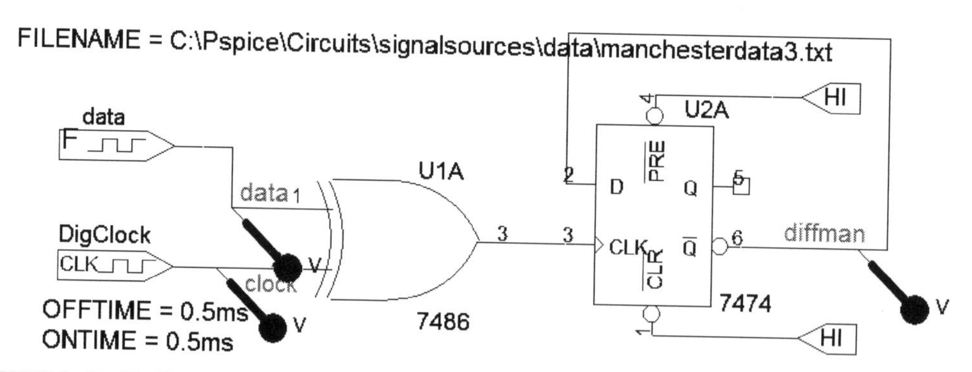

FIGURE 2.17: Differential Manchester coder.

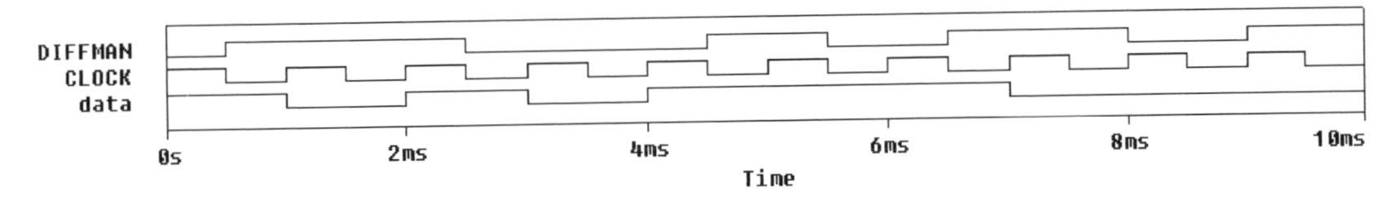

FIGURE 2.18: Differential Manchester signals.

2.5 ALTERNATE MARK INVERSION ENCODING

Alternate mark inversion (AMI) encoding is also known as bipolar return to zero (BPRZ) and is a three-level line code used in the 30-channel TDM PCM E1 system. Return to zero-alternate mark inversion RZ AMI signal has every alternate "one" polarity reversed, with zeros represented as a zero DC level. The alternating characteristic is an advantage because it is easy to recognize a line code violation. AMI production is shown in Fig. 2.19.

The data stimulus for this circuit is provided by a **STIM1** part called **Datagen** and is the same as that used previously where a digital signal pattern is created by looping the signal three times. The alternating mark inversion signal is clearly observed in Fig. 2.20.

The **FFT** icon, when selected, demonstrates that no DC is present in the AMI spectrum in Fig. 2.21.

2.5.1 AMI Decoding

Fig. 2.22 shows the format for the AMI text file inputted using a **VPWL_F_RE_FOREVER** part. The text file is located at C:\Pspice\Circuits\signalsources\data\ami.txt and was created by copying the AMI output signal from the **PROBE** output from the last simulation results.

The schematic in Fig. 2.23 is for recovering data from an AMI encoded signal. This AMI decoder contains a monostable multivibrator with Schmitt-trigger inputs. The components *R2* and *C2*, connected to the supply voltage, set the pulse width in a real circuit but are not

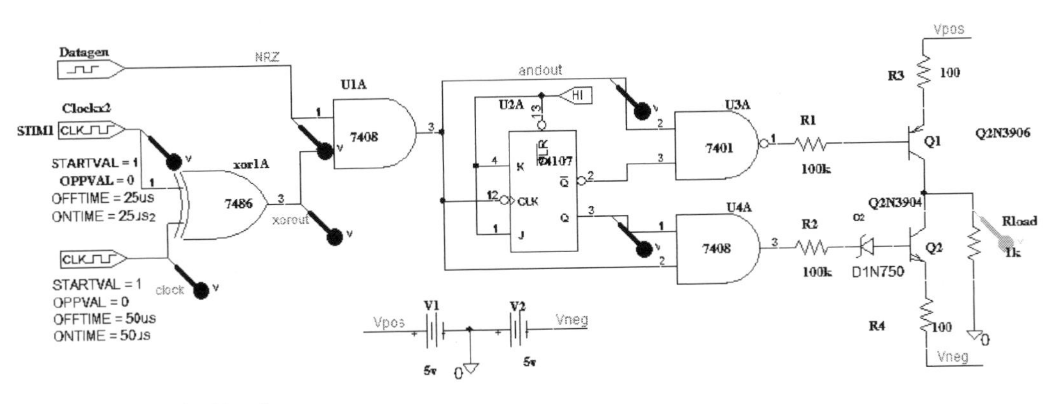

FIGURE 2.19: AMI coder.

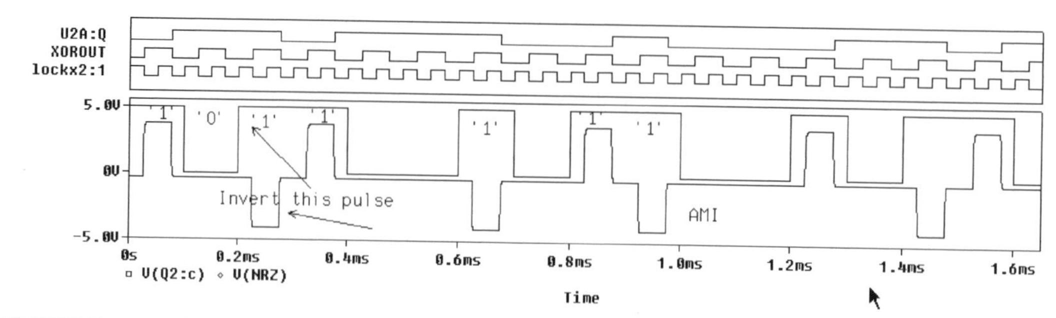

FIGURE 2.20: AMI waveforms.

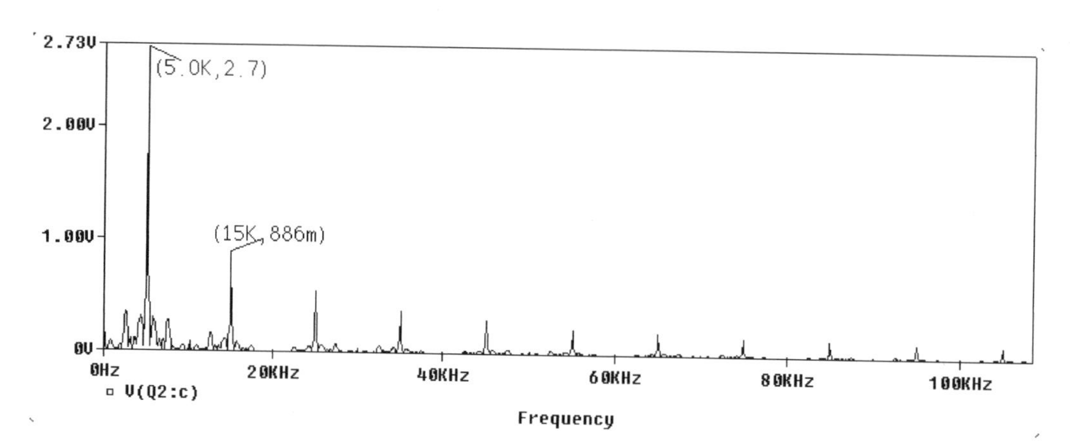

FIGURE 2.21: Spectrum of AMI signal.

T1	T2	T3	T4	T5	T6	T7	T8	T9	T10	V1	V2	V3	V4	V5	V6	V7	V8	V9	V10	Value
0us	50us	50.1us	225us	225.1us	275us	275.1us	325us	325.1us	375us	4v	4v	0v	0v	-4v	-4v	0v	0v	4v	4v	VPWL

FIGURE 2.22: AMI signal parameters.

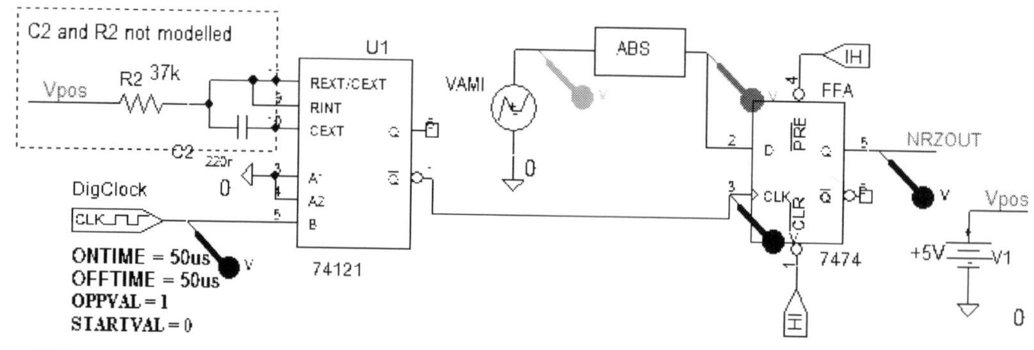

FIGURE 2.23: AMI decoding.

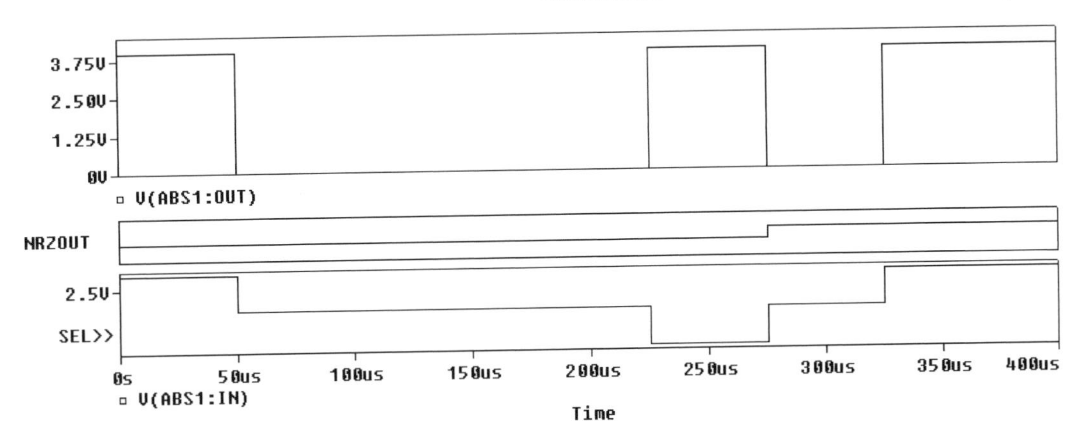

FIGURE 2.24: Decoded AMI signal.

modeled in PSpice. We must set the pulse width by selecting the 74121 IC, **Rclick** and select **Edit Properties.** In the spreadsheet, enter the pulse width in the **PULSE** box (the default value is 30 ns). The **ABS** part converts the bipolar signal into a polar signal.

The recovered NRZ is now shown in Fig. 2.24.

2.6 DUO-BINARY BASEBAND SIGNALING

Duo-binary or partial response signaling is a technique where intersymbol interference (ISI) is added in a controlled manner to the transmitted stream for the purposes of reducing the filter design and the need for Nyquist filtering requirements [ref: 3]. Fig. 2.25 shows a duo-binary

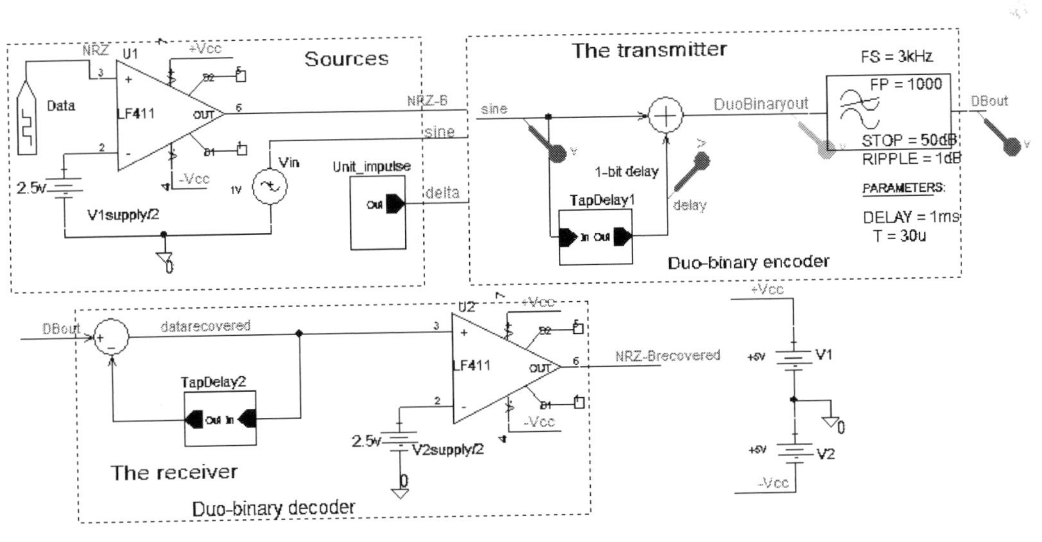

FIGURE 2.25: Duo-binary signaling.

COMMAND1	COMMAND2	COMMAND3	COMMAND4	COMMAND5	COMMAND6	COMMAND7	COMMAND8	COMMAND9	COMMAND10	COMMAND11
LABEL=STARTLOOP	+0ms 1	+1ms 1	+1ms 0	+1ms 0	+1ms 1	+1ms 0	+1ms 1	+1ms 1	+1ms 1	+0ms GOTO STARTLOOP 100 TIMES

FIGURE 2.26: Input test data.

or partial response signaling baseband technique, with a pulse-shaping filter to overcome ISI by introducing a controlled amount of ISI.

The NRZ-B input data is applied using a **STIM1** part called **Data** with parameters as shown in Fig. 2.26.

This data stream is then delayed by an amount equal to the bit period, T_B. The delay is achieved using a correctly-terminated transmission line called a T part [ref: 9 Appendix A]. Correctly-terminating a line means placing a resistance across the input and output terminals whose value is the same as the characteristic impedance. The composite signal from the output of the **SUM** part is then filtered using an ideal pulse-shaping low-pass filter using a **LOPASS** part with a cut-off frequency equal to $0.5 \times 1/T_B$ as shown in Fig. 2.25. The three-level signal in Fig. 2.27 has positive and negative amplitudes that are twice the NRZ-B amplitude.

The transfer function for the duo-binary system, assuming an ideal low-pass filter, is

$$\begin{aligned} H(f) &= H(f)_{\text{ideal}}(1 + e^{-sT}) = H(f)_{\text{ideal}}(1 + e^{-j2\pi f T_b}) \\ &= H(f)_{\text{ideal}}(e^{j\pi f T_b} + e^{-j\pi f T_b})e^{-j\pi f T_b}. \end{aligned} \tag{2.1}$$

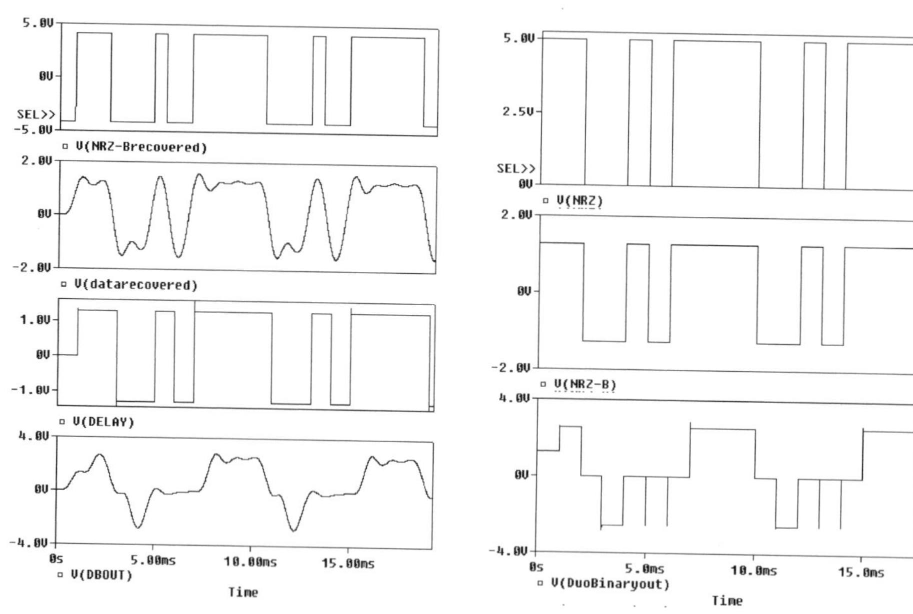

FIGURE 2.27: Duo-binary waveforms.

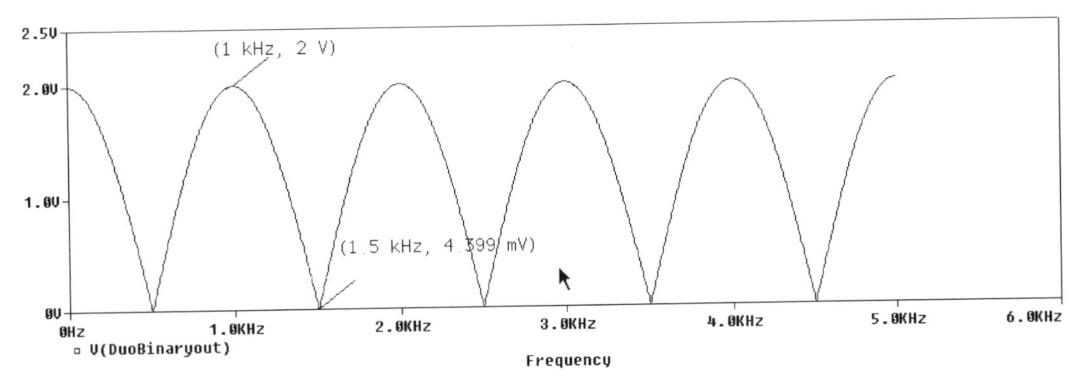

FIGURE 2.28: The duo-binary frequency response.

We may write the overall transfer function by applying Euler's expression to (2.1):

$$H(f) = \begin{cases} 2\cos(\pi f T_b)e^{-j\pi f T_b} \rightarrow |f| \leq 0.5x1/T_b \\ 0 \quad \text{elsewhere.} \end{cases} \tag{2.2}$$

The duo-binary frequency response in Fig. 2.28 was obtained by renaming the input wire segment to sine and changing the simulation profile to AC.

Applying the inverse Fourier transform to equation (2.2), yields $h(t)$ as

$$h(t) = \frac{\sin(\pi t/T_b)}{\pi t/T_b} + \frac{\sin(\pi(t - T_b)/T_b)}{(t - T_b)/T_b}. \tag{2.3}$$

The impulse response is obtained by changing the input signal wire segment to impulse. This results in the display shown in Fig. 2.29.

2.6.1 Use of a Precoder in Duo-Binary Signaling

Fig. 2.30 shows a scheme for utilizing a precoder to overcome the problem of error propagation.

The flip-flops must be initialized to a certain state. Select the **Simulation Setting** menu and then select the **Options** tab. In the **Category** box select **Gate-Level Simulation**. This will then show the **Timing Mode** - select **Typical**. In the **Timing Mode** change **Digital Setup** from **Typical** to **Maximum**. The **Initialize all flip flops** is set to **All 0**. Failing to do this will

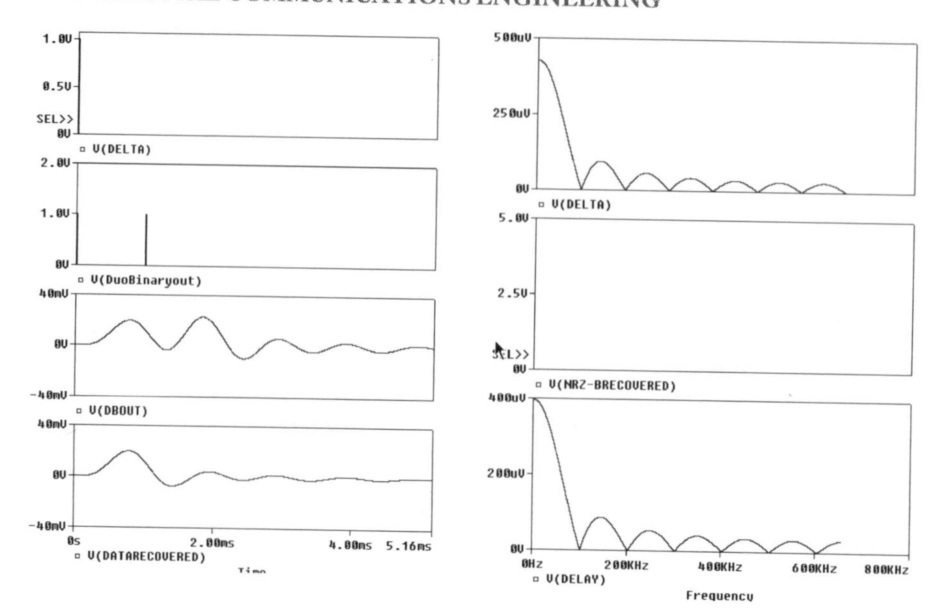

FIGURE 2.29: Impulse response.

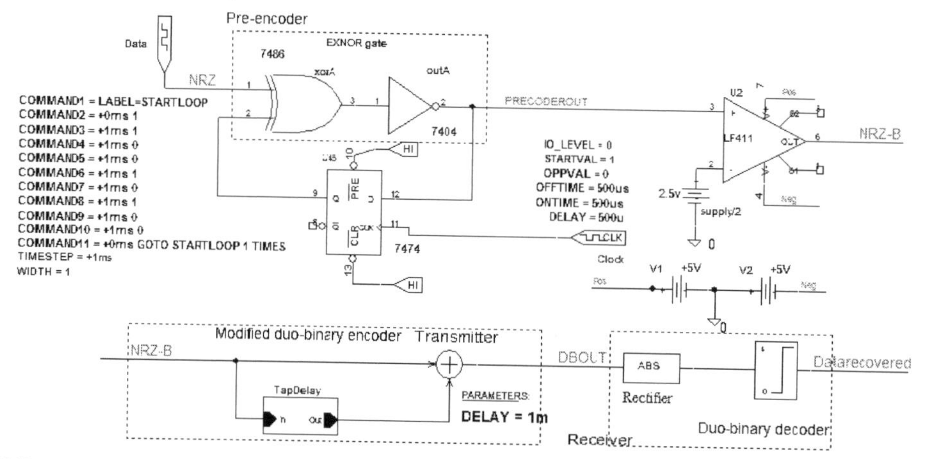

FIGURE 2.30: Precoder.

result in the flip-flop being in an undetermined state, i.e., **All X**, and shows up in the plot as two red lines. The waveforms for this circuit are shown in Fig. 2.31.

2.7 INTEGRATE AND DUMP MATCHED FILTER BASEBAND RECEIVER

The Integrate and Dump matched filter in Fig. 2.32 applies an NRZ-B signal and noise to an integrator. Each bit period is integrated, and bit recovery is possible, provided the noise is random.

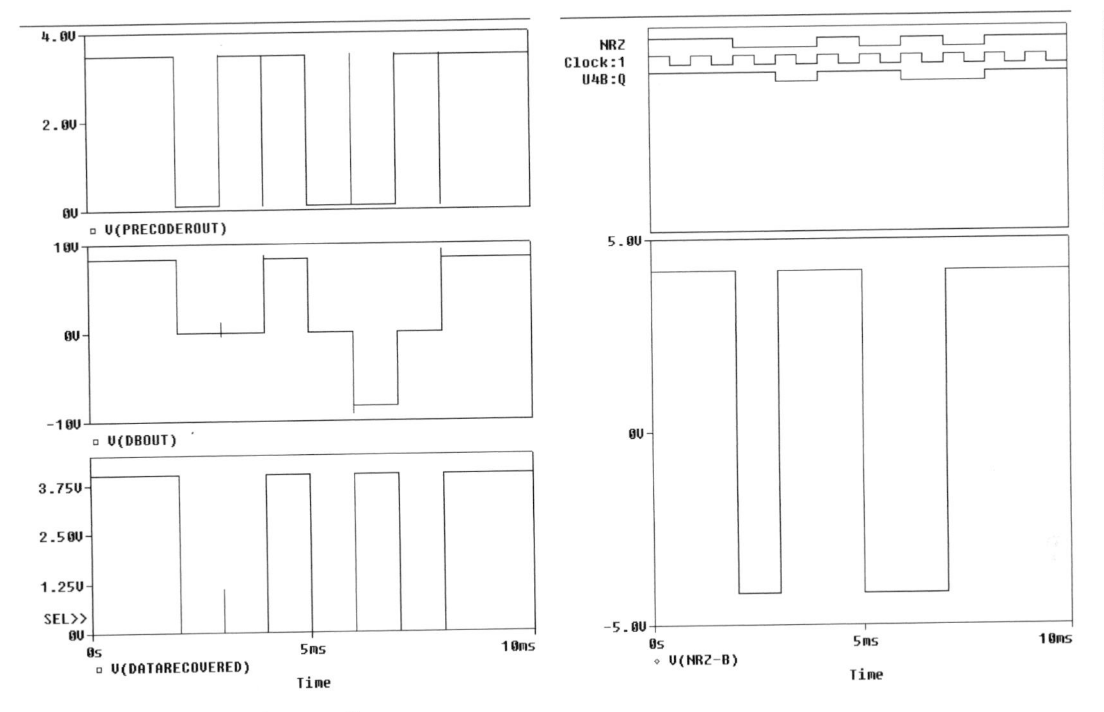

FIGURE 2.31: Precoder waveforms.

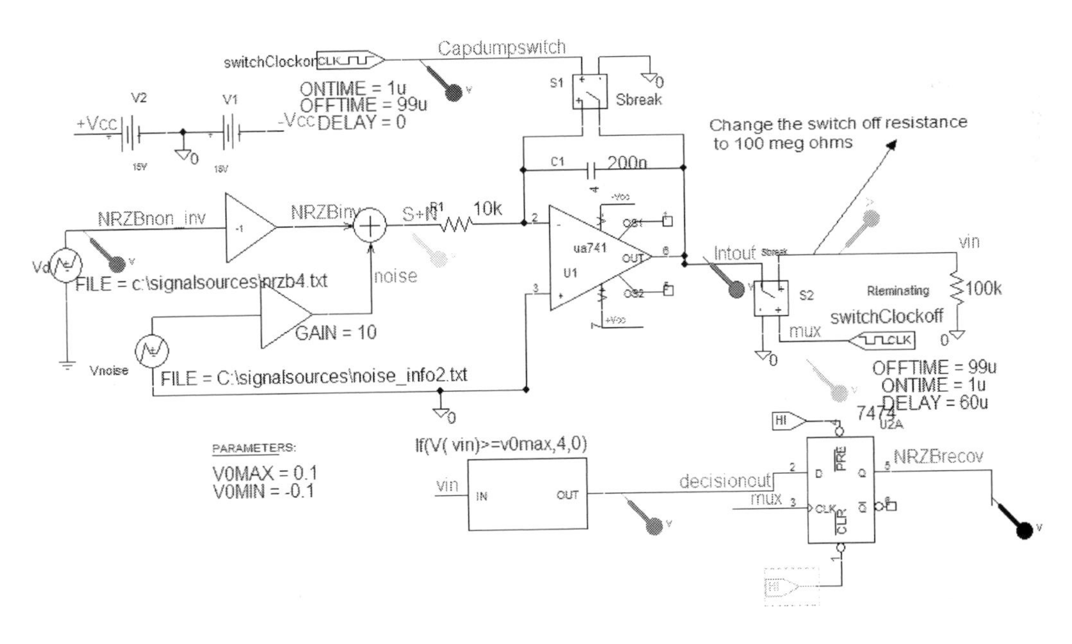

FIGURE 2.32: Integrate and Dump matched filter.

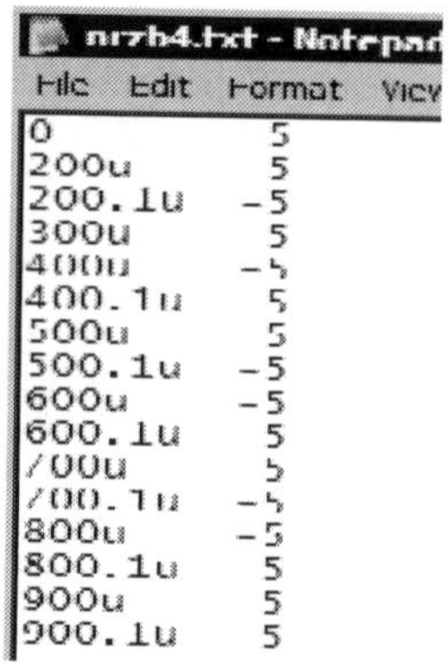

```
nrzb4.txt - Notepad
File  Edit  Format  View
0          5
200u       5
200.1u    -5
300u       5
400u      -5
400.1u     5
500u       5
500.1u    -5
600u      -5
600.1u     5
700u       5
700.1u    -5
800u      -5
800.1u     5
900u       5
900.1u     5
```

FIGURE 2.33: NRZ-B data.

The input NRZ-B data was created by entering values into a text editor (or, alternatively, use the file created by a **Vector** part), as shown in Fig. 2.33. The signal has a period of 100 μs, and the ASCII file is then applied to the circuit using a **FileStim** part. Noise picked up in the transmission path is simulated by applying another ASCII signal using a **VPWL_F_RE_FOREVER** generator part located at the directory C:\signalsources\noise\noise_info2.txt. We vary the noise amplitude using a **GAIN** part, or alternatively, we can use the voltage-scaling factor (VSF—one of the generator parameters). Thus, doubling the VSF value doubles the noise amplitude.

The time constant, $\tau = CR = 2$ms, is much greater than the 100 μs symbol period. The baseband waveforms in Figs. 2.34 and 2.35 show the NRZ-B signal integrated over a symbol period. At the end of each 100 μs period, switch S1 discharges the capacitor. S2 is a sampler switch that operates in the middle of the ramp for 1 μs, thus allowing only 1 μs of the ramp through. The decision threshold circuit is simple using an **ABM1** part and an IF THEN ELSE statement, e.g., If(V(vin) >= v0max, 4, 0). This states that if the input sampled signal is greater than the variable defined in the **PARAM** part, v0max = 100 mV, then the output is 4 V, otherwise it is 0 V.

The D-type flip-flop operates to give a pulse when clocked. Repeat the simulation for different noise amplitudes. Repeat the above exercise but include a passband filter having a 1000 Hz bandwidth.

2.7.1 Example

A 100 baud 1 V NRZ signal contains additive white Gaussian noise (AWGN) and is band-limited to 1000 Hz. This is applied to an Integrate and Dump matched filter and the error probability is estimated by assuming equal probability of a 1 or 0 occurring. The threshold

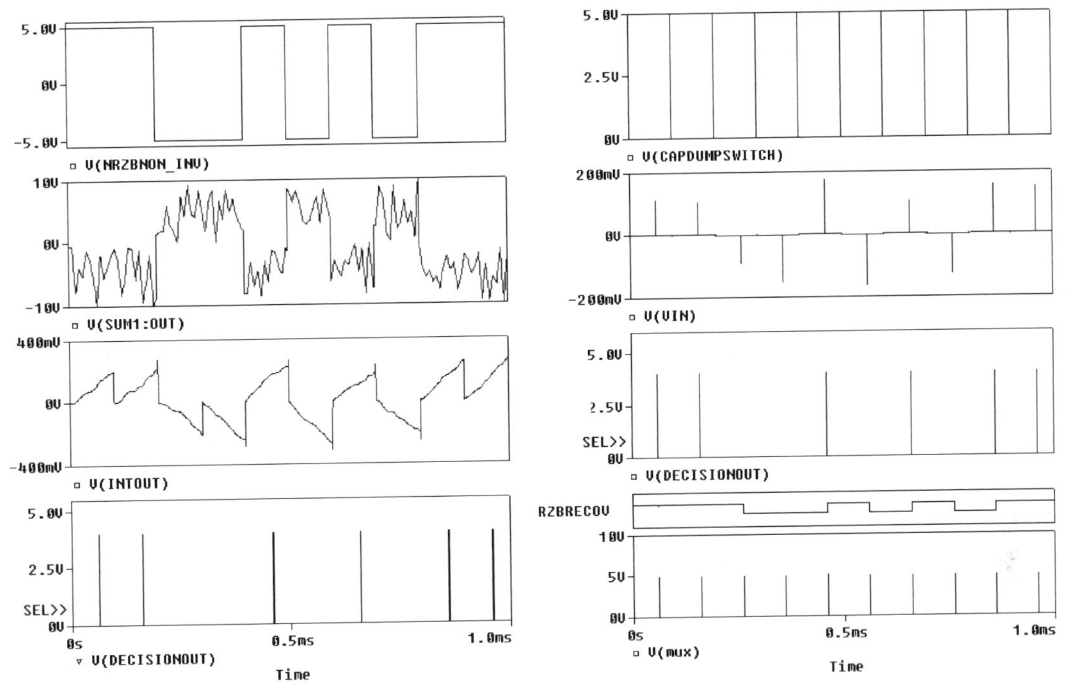

FIGURE 2.34: Integrate and Dump waveforms.

is determined by assuming zero mean noise and with variance $\sigma^2 = 0.125$. The noise variance is

$$\sigma_0^2 = \frac{\sigma^2 T}{2B} = \frac{0.125 \times 0.01}{1000} = 1.25 \times 10^{-6} \Rightarrow \sigma_0 = 1.12 \times 10^{-3}.$$

After integration, the 1 V signal becomes 0.01 (VT) and the decision threshold becomes 0.005 V. Assuming a base probability for 0 and 1 as 0.5, the error probability is

$$0.5\,Q\left(\frac{0.005}{\sigma_0}\right) + 0.5\,Q\left(\frac{0.005}{\sigma_0}\right) = Q\left(\frac{0.005}{1.12 \times 10^{-3}}\right) = Q(4.46) = 3.5 \times 10^{-6}. \quad (2.4)$$

The decision threshold is 0.5 V for a 1 V pulse and noise variance $\sigma^2 = 0.125$ (without the Integrate and Dump circuit). The error probability of a noisy sample with standard deviation of 0.35 exceeding 0.5 V when a zero is transmitted is less than -0.5 V. The bit error rate for a transmitted "1" is calculated as $Q(0.5/0.35) = Q(1.42) \sim 8 \times 10^{-2}$ etc., which is a poor BER rate.

2.8 EXERCISES

1. A modification to the DB system in Fig. 2.35 produces a response with a null at DC, and is useful for channels having a poor low-frequency response.

The modified Duo-binary frequency response is shown in Fig. 2.36.
The modified Duo-binary signals are shown in Figure 2.37.

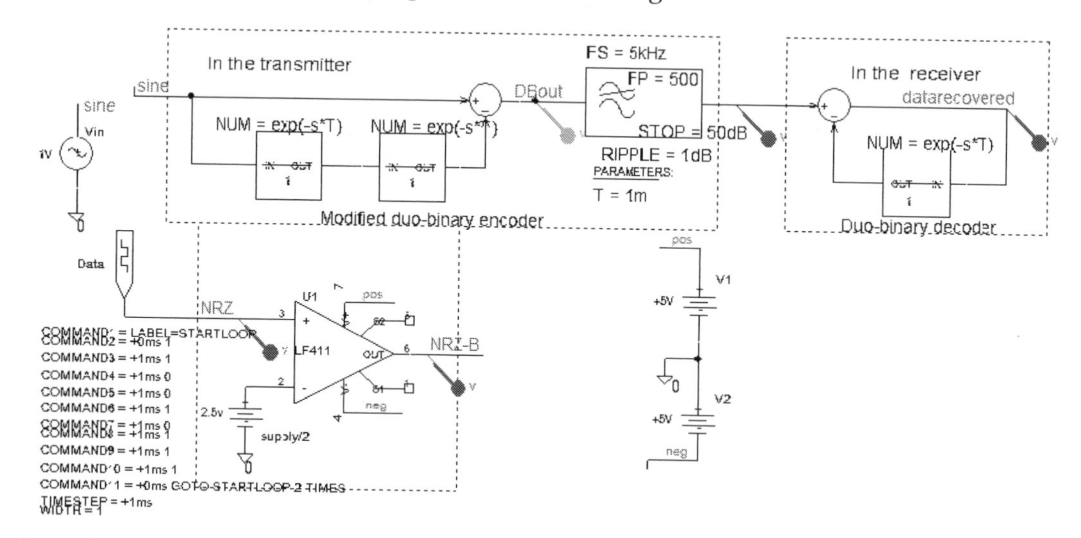

FIGURE 2.35: Modified duo-binary transmitter and receiver.

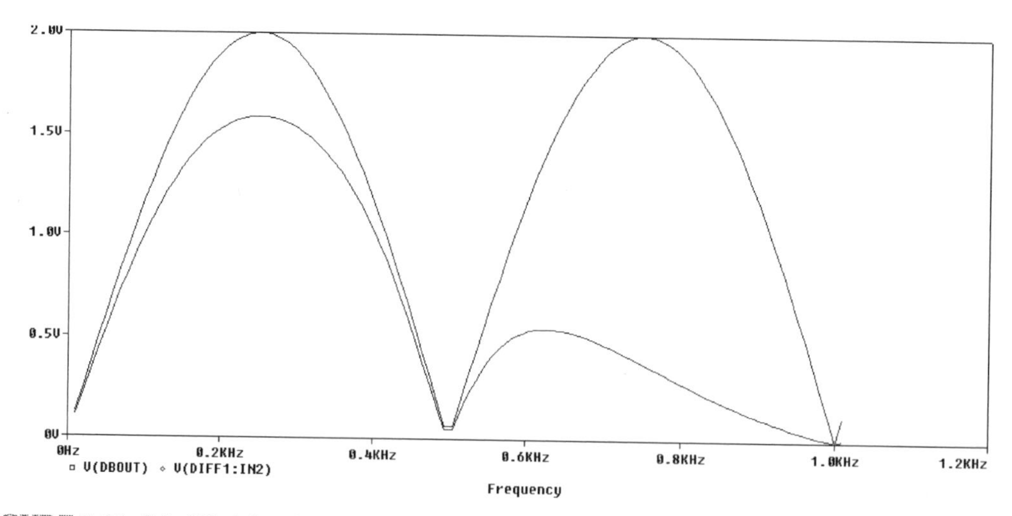

FIGURE 2.36: Modified duo-binary frequency response.

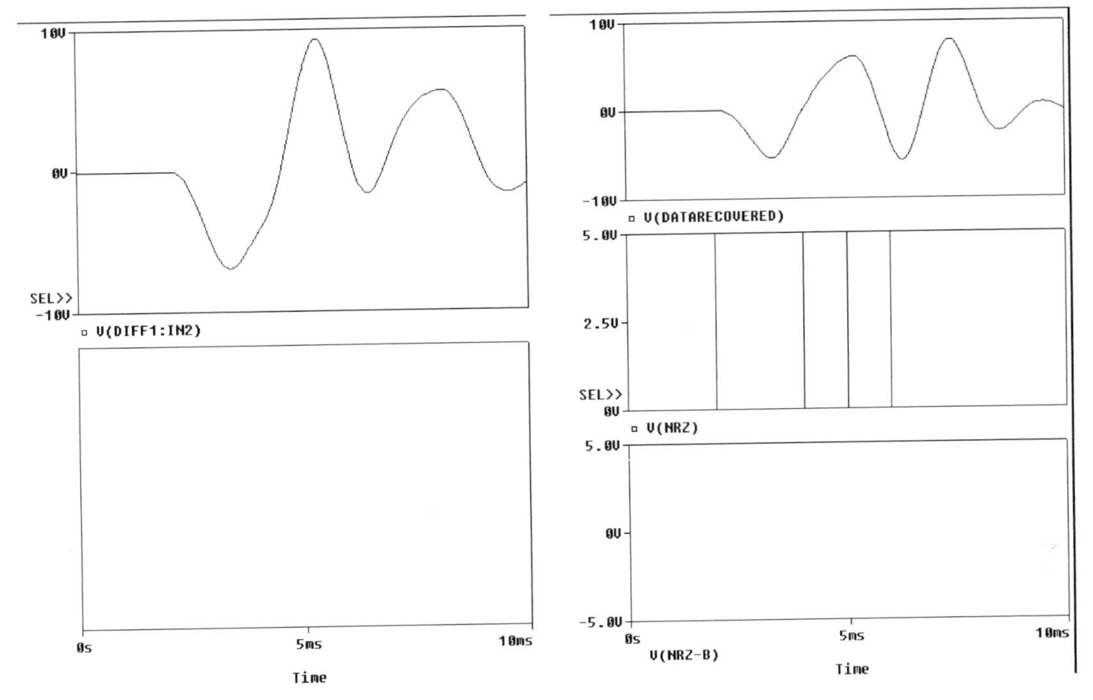

FIGURE 2.37: Modified duo-binary digital signals.

CHAPTER 3
Sampling and Pulse Code Modulation

3.1 SINGLE-CHANNEL PULSE CODE MODULATION

Telephone transmission lines have a bandwidth of 300–3400 Hz (bandwidth of 3.1 kHz). To transmit multiple voice signals simultaneously on the same transmission line requires the signals to be time- or frequency-division multiplexed. In this chapter, we investigate time-division multiplexed pulse code modulation (TDM PCM-Alec Reeves: 1902–1971)), which requires sampling the speech signals at a minimum rate of twice the highest frequency contained in the speech. The speech is filtered with a low-pass filter that has a cut-off frequency of 3400 Hz. Filtering prevents a phenomenon called aliasing where extra frequencies are produced if the sampling is not at the Nyquist rate, i.e., 6800 samples/s (Nyquist sampling theorem) [ref: 8 Appendix A]. This rate is increased to 8000 Hz sampling rate to allow for nonideal filtering in order to recover the signal completely in the receiver. Fig. 3.1 shows a single-channel PCM block diagram.

FIGURE 3.1: Single-channel PCM system.

To limit the frequency spectrum of the speech to a value below half the sampling rate, and prevent aliasing, the analog speech signal is low-pass filtered with a cut-off frequency of 3.4 kHz. This filter is called an antialiasing filter because it prevents aliasing frequencies appearing in the output. The filtered speech is then sampled and companded. Companding is formed from the words compressing/expanding where the signal is first compressed to improve the signal to noise ratio and expanded in the receiver. In PCM systems, the initial signal processing occurs in an integrated circuit called Codec, where the 12-bit digital code is compressed to 8 bits. Each sampled value is assigned one of 256 ($2^8 = 256$) discrete levels for the maximum amplitude range of -1 V to 1 V or 2 V, where V is the largest signal applied and each quantized sample is represented by an 8-bit code.

Encoding a quantization level into 8-bit words produces a single-channel transmission bit rate equal to 8 bits $\times$ 8 kHz $= 64$ kbps. The European E1 primary multiplexing format is the 30-channel 2.048 Mb/s bit rate, whilst the T1 system used in Japan and North America has a 1.5 Mb/s bit rate. The E1 system has 8-bit 32 time slots to accommodate 30 voice signals plus two other time slots for framing, alarm, and signals other than voice information (i.e., ringing tone, engaged tone, call forward, etc.). Time slot 16 accommodates two signaling channels, and hence we need a multiframe comprising 16 frames with a period of 2ms containing 4096 bits.

3.2 COMPANDING CHARACTERISTICS

Linear quantizing is where an equal number of quantized levels are allotted for low- and high-level signals and results in a very poor signal to quantization noise ratio. A nonlinear system increases the number of decision levels for small amplitude signal levels and yields an overall improved quality. In the European E1 system, prior to transmission, the 12-bit digital signal is compressed to 8 bits using a 15-segmented A-law, where the compression parameter is $A = 87.6$. The compressor characteristic is defined for two regions as

$$F_A(x) = \text{sgn}(x)\left[\frac{A\,|x|}{1+\ln(A)}\right] \quad \text{for} \quad 0 \le |x| \le \frac{1}{A}$$

$$= \text{sgn}(x)\left[\frac{87.6\,|x|}{1+\ln(87.6)}\right] \quad \text{for} \quad 0 \le |x| \le 0.0114 \qquad (3.1)$$

$$F_A(x) = \text{sgn}(x)\left[\frac{1+\ln(A\,|x|)}{1+\ln(A)}\right] \quad \text{for} \quad \frac{1}{A} \le |x| \le 1$$

$$= \text{sgn}(x)\left[\frac{1+\ln(87.6\,|x|)}{1+\ln(87.6)}\right] \quad \text{for} \quad 0.0114 \le |x| \le 1. \qquad (3.2)$$

The signum function is defined as

$$\text{sgn}(x) = \begin{cases} -1 \text{ for } x < 0 \\ 1 \text{ for } x > 1, \end{cases}$$

and is zero when x is zero. The **LIMIT** part defines the signal constraints in (3.1) and (3.2). Note that **LOG()** in **PROBE** is natural log, whereas **LOG10()** is log to the base 10. The continuous A-law companding function is simulated by entering the following expression into the **ABM1** part using an IF THEN ELSE statement to manage the input level constraints as if(V(%IN)<=1/A,sgn(V(%IN))*A*abs(V(%IN))/(1+log(A)),sgn(V(%IN))*(1+log(A*abs×(V(%IN))))/(1+log(A))).

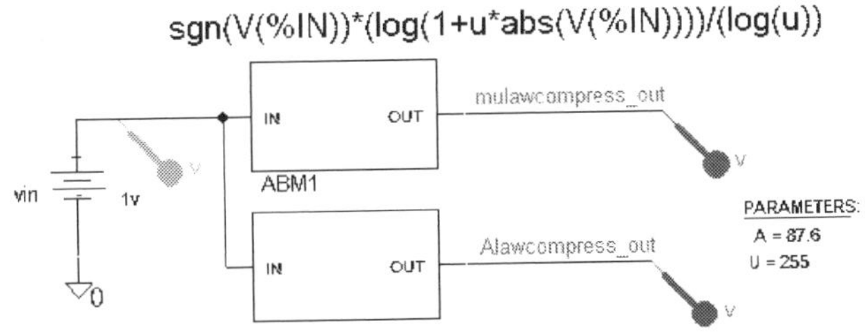

FIGURE 3.2: Modeling the μ- and *A*-law characteristic.

The μ-law characteristic is defined as

$$F_u(x) = \operatorname{sgn}(x)\left(\frac{1+\ln(255\,|x|)}{\ln(1+255)}\right) \quad \text{for} \quad 0 \leq |x| \leq 1,$$

where $u = 255$. This is entered in the **ABM1** part as sgn(V(%IN))*(log(1+u*abs(V(%IN))))/(log(u)). Sweep the input voltage as shown in Fig. 3.3. Set **Analysis Type: DC Sweep, Sweep Variable** = Voltage source, **Name**: vin, **Linear, Start Value** = 0, **End Value** = 1, **Increment** = 0.0001. The *A*-law characteristic shows higher amplitude input signals compressed.

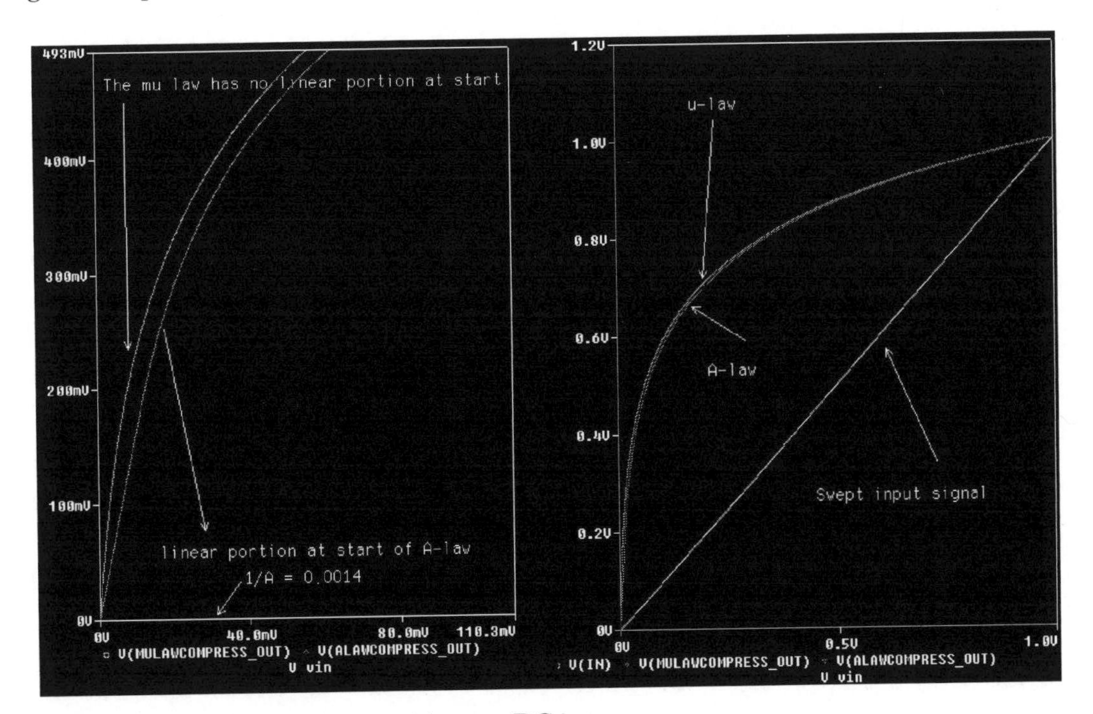

FIGURE 3.3: *A*-law characteristic with swept DC input.

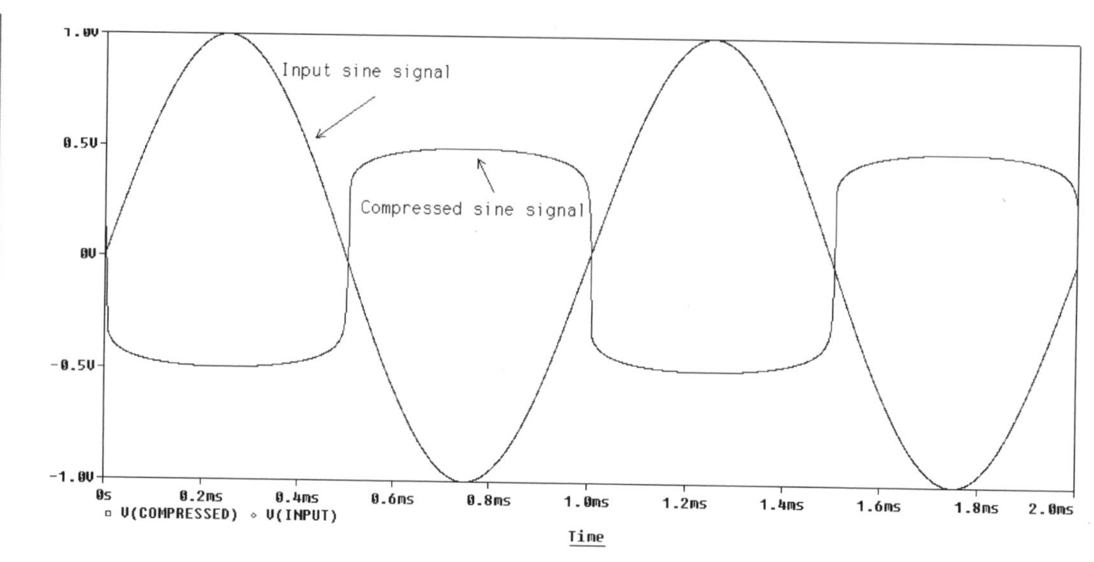

FIGURE 3.4: Sinusoidal input signal.

This nonlinear companded PCM (CPCM) uses a fixed number of quantized levels, but with a much higher percentage of the levels assigned to smaller amplitude input voltages. A larger signal will have a smaller number of quantizing levels, but this is OK since statistically these signals are not significant. At the receiver, there is an identical nonlinear expander. "Compression and expansion" is known as companding. There are two segmented nonlinear instantaneous companding systems: *A*-law for the European PCM system and μ-law for the American system. Replace the VDC source with a VSIN part and observe the effect on the signal in Fig. 3.4.

The complete μ-law companding characteristic is simulated with the following transient parameters: **Analysis Type**: **DC Sweep, Sweep Variable** = Voltage source, **Name**: vin, **Linear, Start Value** = −1, **End Value** = 1, **Increment** = 0.0001. The complete μ-law characteristic is shown in Fig. 3.5. In practice, however, the compression is not done at the analog level but on a 12-bit digital signal, where it is compressed to an 8-bit format using a segmented *A*-law characteristic. We may use a **Table** part or the **Value List** in the **DC Sweep** menu to achieve the segmentation or chords. In the latter case, tick **Value List** in the **Sweep Type** menu and enter in the **Values** box the following values: 0, 0.0156, 0.0313, 0.0625, 0.125, 0.25, 0.51. The segmented *A*- and μ-law characteristics are shown in Fig. 3.5 with the complete μ-law characteristic shown in the left panel along with swept input signal.

The first bit of the eight bits represents the signal sign. The next three bits ($2^3 = 8$) represent the segment the signal lies in, and the last four bits are for each of the $2^4 = 16$ decision levels where bits are transmitted serially, sign bit first. The dynamic range for the

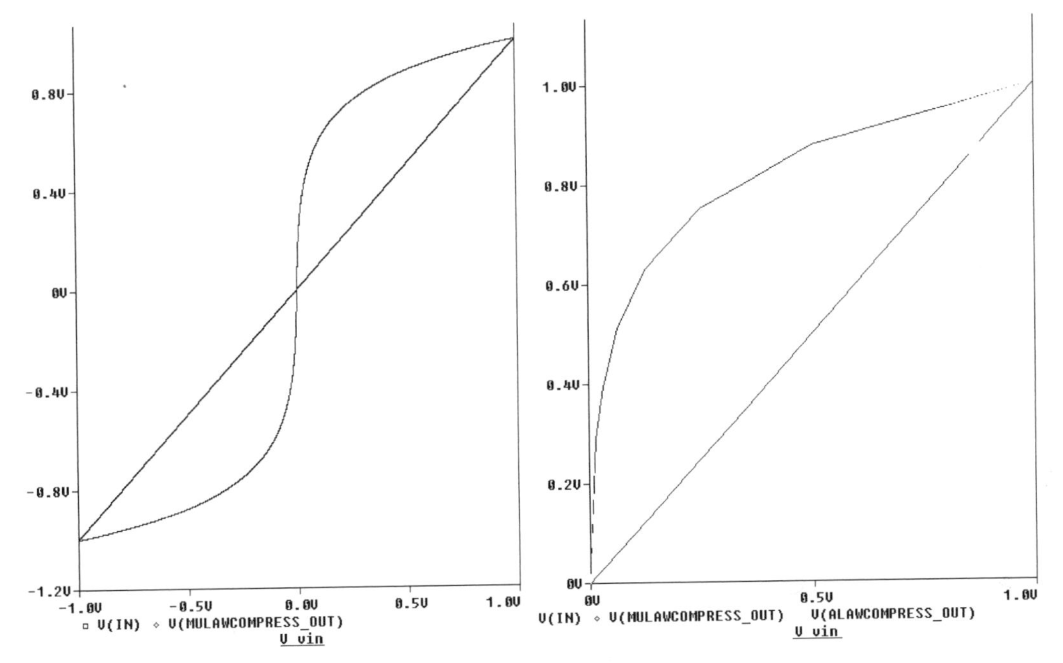

FIGURE 3.5: Segmented *A*-law characteristic.

A-law is $20\log_{10}(4096/15) = 48.7$ dB, where 0–15 spans the first chord. The dynamic range for the µ-law is $20\log_{10}(8159/31) = 48.4$ dB, 0–31 spans the first chord.

3.3 SAMPLING

To recover an analog signal from a sampled signal, and with no aliasing (extra frequencies generated), requires that the signal be sampled in the first place by a sampling frequency that is greater than twice the highest frequency component in the analog signal. A 1 kHz analog signal must be sampled at a rate at least equal to 2 kHz (the work of C. Shannon, 1948), to preserve and recover the waveform exactly. Sampling a signal at a rate below twice its highest frequency produces aliasing components, which are extra aliasing frequencies in the recovered signal. Low-pass filtering the signal prior to sampling prevents aliasing in a fixed sampling system. The sampling rate in PCM telecommunication systems is 8 kHz (period $T_P = 125$ µs), which means a 4 kHz theoretical maximum input signal frequency.

An ideal brick-wall filter that cuts off abruptly at 4 kHz is not practical, so the input signal frequency is limited to 3.4 kHz with a finite filter transition region width. A square wave using a **VPULSE** part is applied to a second-order Sallen and Key active low-pass filter in Fig. 3.6. The square wave simulates a complex signal such as speech because the square wave contains harmonics from DC to infinity. The filter output is then sampled using a dual sampling switch IC CD 4016.

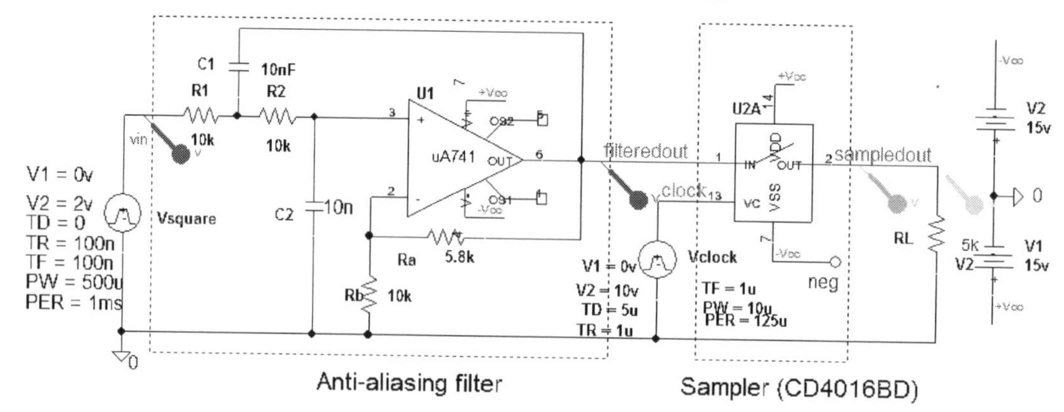

FIGURE 3.6: Antialiasing filter and sampler circuit.

Design a second-order antialiasing Sallen and Key LPF filter to extract the fundamental harmonic of the square wave, and attenuate higher order harmonics. The filter ensures that frequencies greater than twice the sampling frequencies are sufficiently attenuated and thus eliminates alias components.

3.3.1 Sallen and Key Antialiasing Active Filter

The antialiasing second-order Butterworth Sallen and Key active filter is designed to meet the specification $A_{\max}$, $A_{\min}$, ω_c, and ω_s. These are the passband and stopband gains, passband and stopband edge frequencies, respectively. The passband cut-off frequency is $\omega_c = 1/C_1 R_1$ and the gain in the passband region is $20\log(1 + R_a/R_b)$ with the roll-off rate of -40 dB per decade. We need to extract the fundamental harmonic (the 1 kHz fundamental harmonic component) from the square wave. The following analysis yields a low-pass transfer function, which produces maximum attenuation of 1 dB at the fundamental component of 1 kHz. The 3 kHz, or third harmonic, should be attenuated by 12 dB [ref: 1 Appendix A]. The frequency correction factor is calculated:

$$\varepsilon = \sqrt{(10^{0.1.A_{\max}} - 1)} = \sqrt{(10^{0.1.1} - 1)} = 0.508. \qquad (3.3)$$

The filter order is calculated using the following expression:

$$n = \frac{\log_{10}\left[\frac{(10^{0.1A_{\min}}-1)}{(10^{0.1A_{\max}}-1)}\right]}{2\log_{10}\left(\frac{\omega_s}{\omega_p}\right)} = \frac{\log_{10}\left[\frac{(10^{0.1.12}-1)}{(10^{0.1.1}-1)}\right]}{2\log_{10}\left(\frac{2\pi 3000}{2\pi 1000}\right)} = 1.84 \approx 2. \qquad (3.4)$$

A normalized second-order Butterworth approximation loss function is:

$$A(\$) = \$^2 + 1.414\$ + 1. \qquad (3.5)$$

This function has a normalized cut-off frequency of 1 r/s, that is denormalized by replacing $ with

$$\$ = s\frac{\varepsilon^{1/n}}{\omega_c} = s\frac{0.508^{1/2}}{2\pi 1000} = s\frac{0.712}{6283}. \quad (3.6)$$

Substitute this value into (3.5) and invert to yield the transfer function:

$$H(\$)\big|_{\$=s\frac{0.712}{6283}} = H(s) = \frac{1}{(s\frac{0.712}{6283})^2 + 1.414(s\frac{0.712}{6283}) + 1} x \frac{(6283/0.712)^2}{(6283/0.712)^2}$$

$$= \frac{7.787.10^7}{s^2 + 1.247.10^4 s + 7.787.10^7}. \quad (3.7)$$

We are now in a position to obtain component values by comparing the denominator transfer function coefficients of (3.7) to the standard second-order transfer function denominator $s^2 + \omega_p/Q + \omega_p^2$ and the transfer function is

$$\frac{E_0}{E_i} = \frac{k/C^2R^2}{s^2 + s(3-k)/CR + 1/C^2R^2}. \quad (3.8)$$

Consider the non-s coefficient term

$$\omega_p^2 = 7.787 \times 10^7 = \frac{1}{(CR)^2} \Rightarrow \omega_p = 8824.4 = \frac{1}{CR}. \quad (3.9)$$

The capacitance is calculated for $R = 100\ \text{k}\Omega$ as

$$C = \frac{1}{8824.4 \times 10^5} = 1.139\ \text{nF}. \quad (3.10)$$

Compare the s coefficient terms in the denominator of (3.8) and (3.7), and substitute the values for C and R:

$$\frac{\omega_p}{Q} = \frac{3-k}{CR} = 1.247 \times 10^4 \Rightarrow 3 - k = 1.247 \times 10^4 \times 1.139 \times 10^{-9} \times 10^5 = 1.42. \quad (3.11)$$

The passband gain is

$$k = 3 - 1.42 = 1.58 = 1 + R_a/R_b \Rightarrow R_a/R_b = 0.58 or R_a = 0.58R_b. \quad (3.12)$$

If $R_b = 10\ \text{k}\Omega$, then $R_a = 5.8\ \text{k}\Omega$ and a gain of 1.58. Set **Output File Options/Print values in the output file** to 1 ms, **Run to time** to 10 ms, **Maximum step size** = 10 u, and press the **F11** key to simulate. The frequency spectrum for a 1 kHz square wave input signal and other signals are shown in Fig. 3.7.

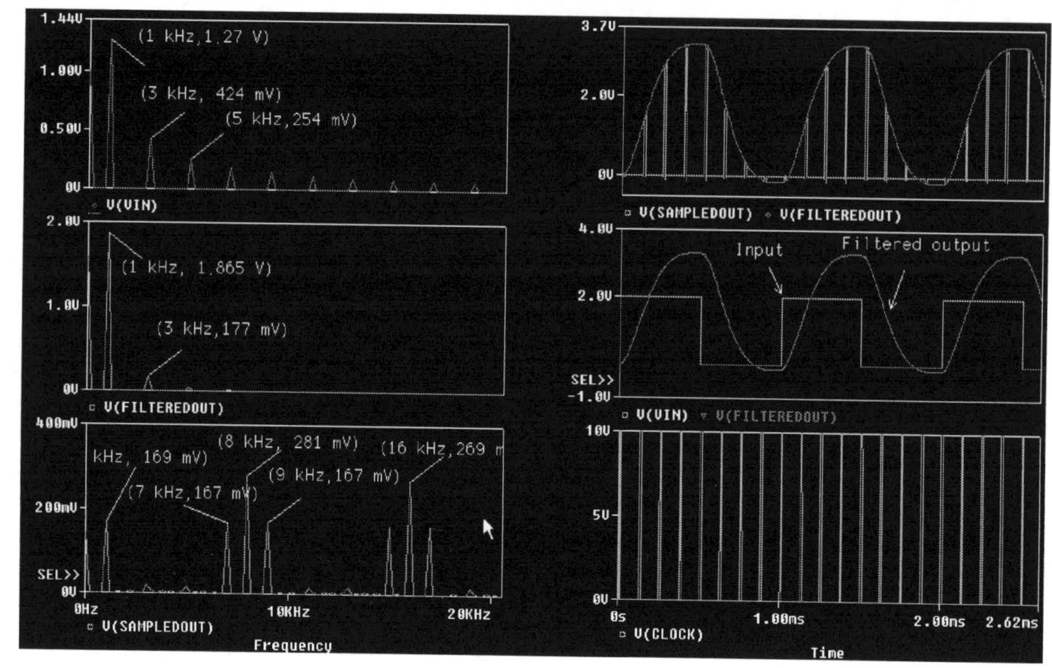

FIGURE 3.7: A 1 kHz square wave spectrum.

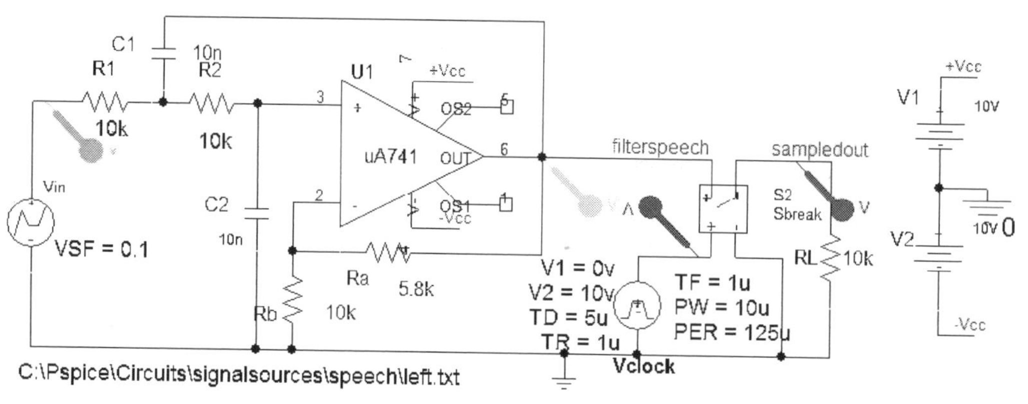

FIGURE 3.8: Input speech file.

3.3.2 Speech Signals

A speech signal is applied using the **VPWL_F_RE_FOREVER** generator as shown in Fig. 3.8.

This generator reads in a speech file "speech.txt," created and stored on the hard drive in the "signalsources" directory. Sampling uses the **Sbreak** switch part operated by a **DigClock** part. Set **Output File Options/Print values in the output file** to 0.1s, **Run to time** to 1s, and **Maximum step size** = 100 u and press the **F11** key to simulate. Fig. 3.9 shows the input and

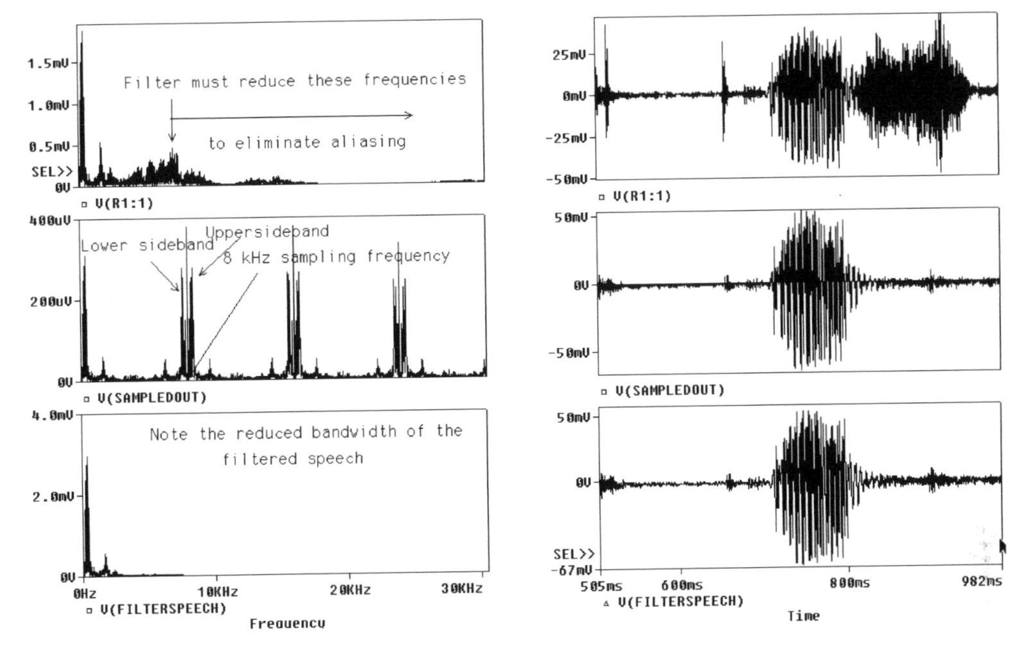

FIGURE 3.9: Speech waveforms.

output speech signals in the time domain, but to see the effect of filtering we need to look at the signals in the frequency domain. Note how the sampled speech sidebands are now located at the sampling frequency, and at multiples of the sampling frequency. The reduced frequency range is evident from the middle trace, which means that the aliasing components are reduced to such a small value that effectively, they are zero.

3.3.3 Sample and Hold

The schematic in Fig. 3.10 investigates how a signal sample is held at a certain level in between sampling times. An ideal **opamp** part is used instead of a ua741 operational amplifier in order

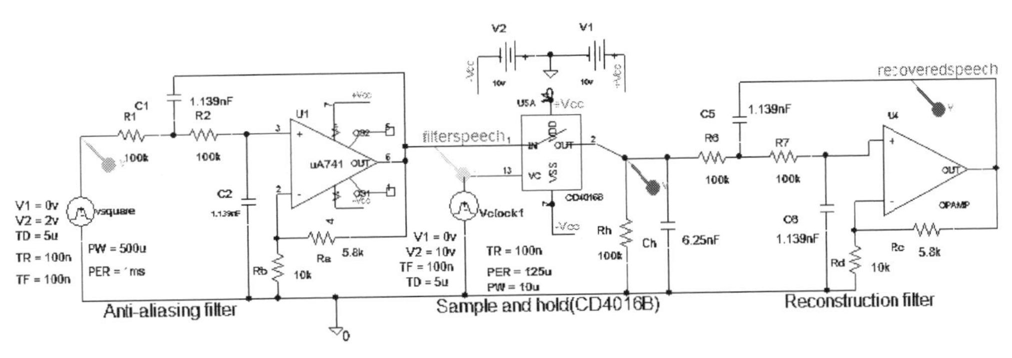

FIGURE 3.10: Sampler with reconstruction filter.

to satisfy the evaluation version limits. Set **Output File Options/Print values in the output file** to 100 ns, **Run to time** to 10 ms, and **Maximum step size** = 10 u and press the **F11** key to simulate and observe, in the right pane of Figure 3.13, the effect of not buffering the sample and hold circuit components.

A square wave signal has a spectrum that extends to infinity and may be used to simulate complex signals such as speech. The sampling frequency in modern telecommunication systems is 8 kHz, which is a sampling period of $T = 125$ μs. This means that the maximum frequency of the input signal should not be greater than 4 kHz. This is achieved by band-limiting the input analog signal to 4 kHz using an active low-pass filter. The antialiasing filter specification defines $A_{\max}$, $A_{\min}$, ω_c, and ω_s as the passband gain, stopband gain, the passband edge frequency, and the stopband edge frequency, respectively. The passband cut-off frequency is $\omega_c = 1/C_1 R$ and the passband gain is $20 \log(1 + R_a/R_b)$. We may investigate aliasing by reducing the sampling rate below the minimum rate required, which for a 3.4 kHz modulating signal is 6.8 kHz and lower.

The undersampled signal spectrum is compared to the correct sample spectrum. Undersampling the input signal will produce aliasing components in the recovered spectrum and contains extra aliasing frequency components. This manifests itself as distortion in the recovered signal. Increase the sampling rate to 20 kHz and observe any changes in the recovered signal. Oversampling reduces the "burden" on the antialiasing filter and thus a simpler filter design would suffice. In the sample and hold circuit, the hold capacitor, C_h, across the hold resistance, R_h, "holds" the instantaneous amplitude of the sampled signal over the sampling period until the next sample and thus produces less distortion in the recovered signal. We may observe the effect of not using a hold circuit by observing the frequency spectrum and measuring the distortion. Sample and hold may introduce aperture error, however, depending on how much the signal changes during the period of sampling. We need to calculate the slope of the 1 kHz sinusoidal signal and multiply it by the sampling period, $T_s = 125$ us, to calculate the maximum amount of error introduced. Consider the instantaneous value of a sinusoidal to be sampled:

$$v_c(t) = V_c \sin \omega t = 1 \sin 2\pi 10^3 t \text{ V}. \tag{3.13}$$

We determine the maximum voltage error from the maximum rate of change, or slope, which occurs at the zero crossover point for the sine wave shown in Fig. 3.11. Differentiating (3.13) yields

$$\frac{dv_c(t)}{dt} = \omega V_c \cos \omega t = 2\pi f \cos \omega t = \frac{\Delta V_c}{\Delta t}. \tag{3.14}$$

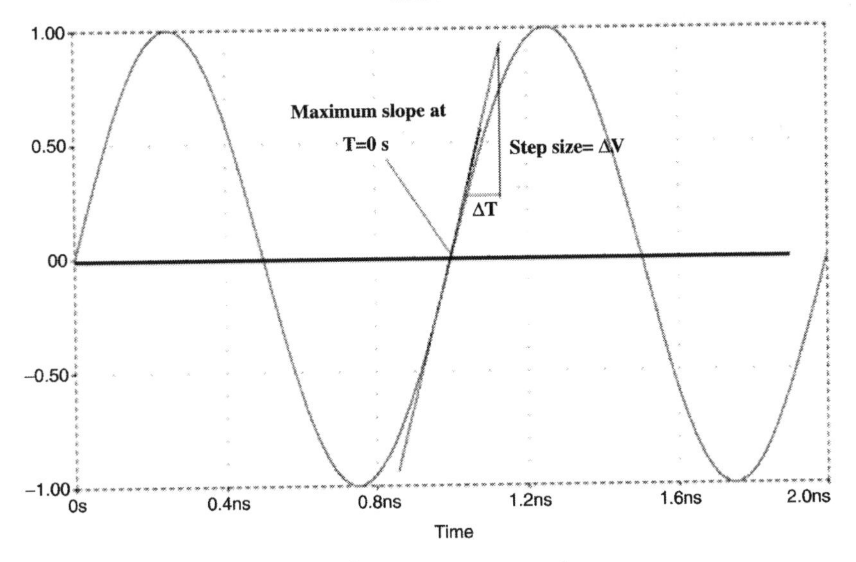

FIGURE 3.11: Maximum slope occurs at the zero crossover point.

The maximum slope occurs at time $t = 0$ s:

$$\left.\frac{dv_c(t)}{dt}\right|_{t=0} = \omega V_c = 2\pi f V_c = \frac{\Delta V_c}{\Delta t} \Rightarrow \Delta V_c = 2\pi f V_c \Delta t. \tag{3.15}$$

Thus, the maximum voltage error is $\Delta V_c = 2\pi f V_c \Delta t$, where

$$2\Delta t = T_{\text{clock}} = \frac{1}{f_{\text{clock}}} \Rightarrow \Delta t = \frac{1}{2 f_{\text{clock}}} = \frac{1}{2 \times 8000 \times 10^3} = 62.5 \times 10^{-6} \text{ s}. \tag{3.16}$$

Substituting this value into (3.15) yields the maximum voltage error as

$$\Delta V_c = 2\pi \times 10^3\, 62.5 \times 10^{-6} = 392 \times 10^{-3} \text{ V}. \tag{3.17}$$

Draw the sample and hold schematic shown in Fig. 3.12. A reconstruction filter is included and has the same cut-off frequency as the input LPF in the sampler circuit considered previously.

The recovered signals in the right panel in Fig. 3.13 show the effect of not buffering the hold circuit.

The sample and hold pulse is the weighted inverted delayed unit step subtracted from a unit step also weighted by V:

$$h(t) = Vu(t) - Vu(t - T_s). \tag{3.18}$$

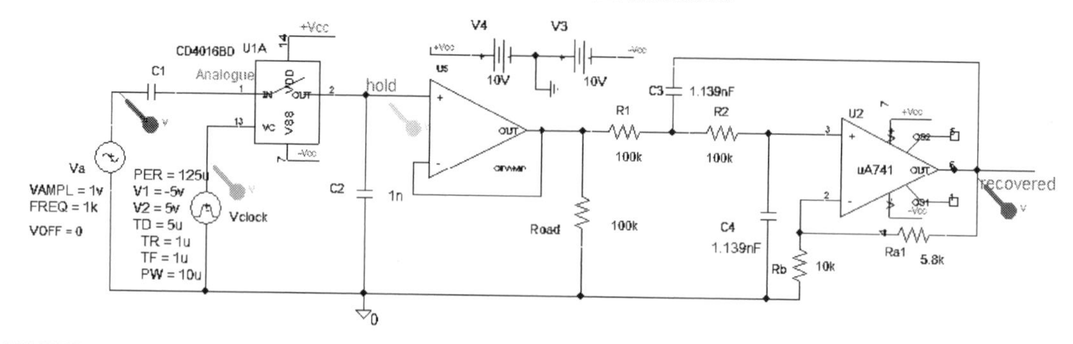

FIGURE 3.12: Sample and hold and recovery schematic.

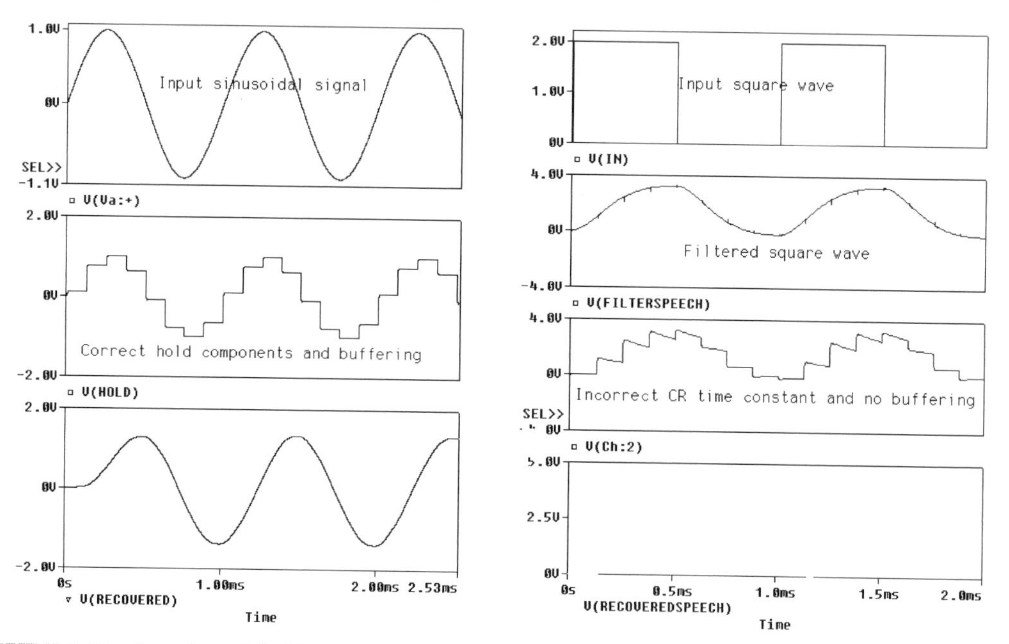

FIGURE 3.13: Sample and hold time signals.

From the inverse Laplace Transform tables, we get an expression in time for the signals in the charging period:

$$F(s) = \int_0^t Ve^{-st}dt = \left[\frac{-V}{s}e^{-st}\right]_0^t = -\frac{V}{s}\left[e^{-t} - e^{-0}\right] = \frac{V}{s}[1 - e^{-t}]. \qquad (3.19)$$

The sampling and hold signals are shown in Fig. 3.14 but note the charging and discharging periods.

A correct choice for the hold circuit components must be chosen; otherwise we get distortion in the final output signal.

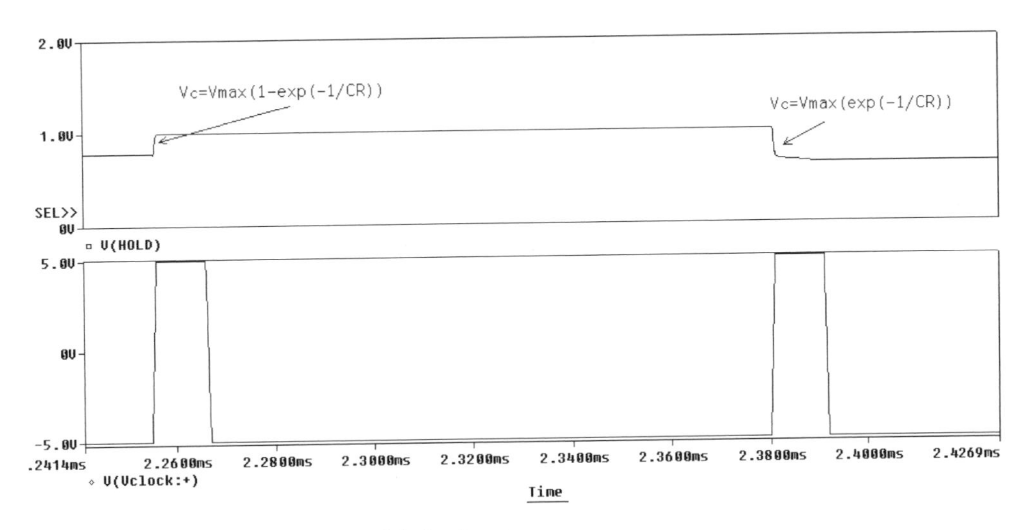

FIGURE 3.14: Magnified sample and hold signals.

3.4 QUANTIZATION NOISE

Quantization noise, q, is an unwanted product produced in the conversion from analog signals into a digital format. Analog to digital converters (ADC) produce this noise and it is reduced by increasing the number of levels, M, in the ADC process. The signal to noise ratio (SNR) in analog circuits is defined as the ratio of the signal amplitude to the noise amplitude present. In sampled systems, however, the signal to quantization noise ratio (SQNR) is the ratio of signal power to quantizing noise power. The easiest way to examine SQNR is to apply a sinusoidal signal $V \cos \omega t$ that occupies the full range of the quantizer. The full-scale value is $V_{\text{FS}} = 2^n q/2$, where n is the number of bits used. The average signal power for full-scale voltage V_{FS} is

$$P_{\max} = \frac{V_{\text{rms}}}{R} = \frac{(V_{\text{FS}}/\sqrt{2})^2}{R} = \frac{(qM/\sqrt{2})^2}{R} = \frac{q^2(2^n/2\sqrt{2})^2}{R} = \frac{q^2 2^{2n}}{8R}. \tag{3.20}$$

The quantizing noise voltage is limited to $\pm\, q/2$ and, except at the signal peak level, the voltage has all values within this range occurring with equal probability. The averaged normalized noise power is determined by integrating the quantization noise voltage (saw-toothed in shape and with a negative slope $-qt/T_0$ over a period T_0):

$$N_q = \frac{1}{T_0}\int_{-T_0/2}^{T_0/2}\left[-\frac{q}{T_0}t\right]^2 dt = \frac{q^2}{3T_0^3}\left[\frac{T_0^3}{8}+\frac{T_0^3}{8}\right] = \frac{q^2}{12}. \tag{3.21}$$

The ratio of the signal power to noise power SQNR for a signal that occupies the full scale is

$$\mathrm{SNQR} = 10\log_{10}\frac{P_{\max}}{P_N} = 10\log_{10}\left[\frac{2^{2n}q^2/8}{q^2/12}\right] = 10\log_{10}\left[\frac{3}{2}2^{2n}\right] = 10\log_{10}\left[\frac{3}{2}M^2\right] \tag{3.22}$$

$$\mathrm{SNQR} = 10\log_{10}2^{2n} + 10\log_{10}1.5 = 20n\log_{10}2 + 10\log_{10}1.5 = 6.02n + 1.77\ \mathrm{dB}. \tag{3.23}$$

However, the SQNR decreases when the input sinusoid is below the full scale of the ADC. Signal power is a function of waveform shape but noise power is independent of the signal shape. The signal power for a sinusoid with amplitude $\pm A$ is $A^2/2$ and A^2 for a square wave and $A^2/3$ for a triangular wave.

3.5 ANALOG TO DIGITAL CONVERSION

An 8-bit analog to digital converter (ADC) and digital to analog conversion (DAC) is shown in Fig. 3.15. The low-pass filter on the output reconstructs the original input signal. To measure the quantization noise we use a notch filter to remove the desired signal leaving the quantization noise. The 8-bit ADC binary output, set to the nearest integer, in given by

$$\frac{V_{\mathrm{in}}}{V_{\mathrm{ref}}}2^{n\mathrm{bits}}. \tag{3.24}$$

The digital to analog converter (DAC) output produces a rough version of the original input signal. A low-pass filter attenuates the high-frequency components leaving the original signal. Consider the input and reference voltages for an 8-bit system. The **VSIN** part parameters are as follows: **VOFF** = 10 V and **VAMPL** = 10 V, giving a peak input of Vin = 20 V. We set the reference to 256 V in order to read the output in binary directly. The binary equivalent for the ADC reference voltage, **VREF**, is equal to 256 V and is ($20 \times 2^8/256$) equal to 10100. From the **PROBE** plot read the most significant bit to the least significant as b4 b3 b2 b1 b0.

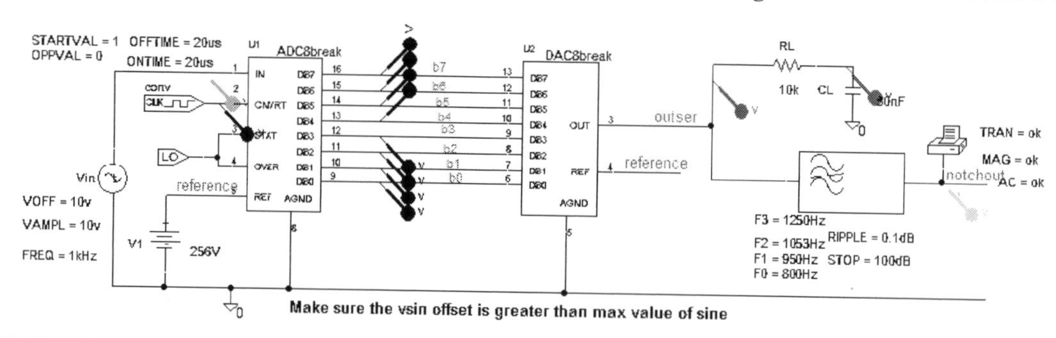

FIGURE 3.15: PCM coder and decoder system.

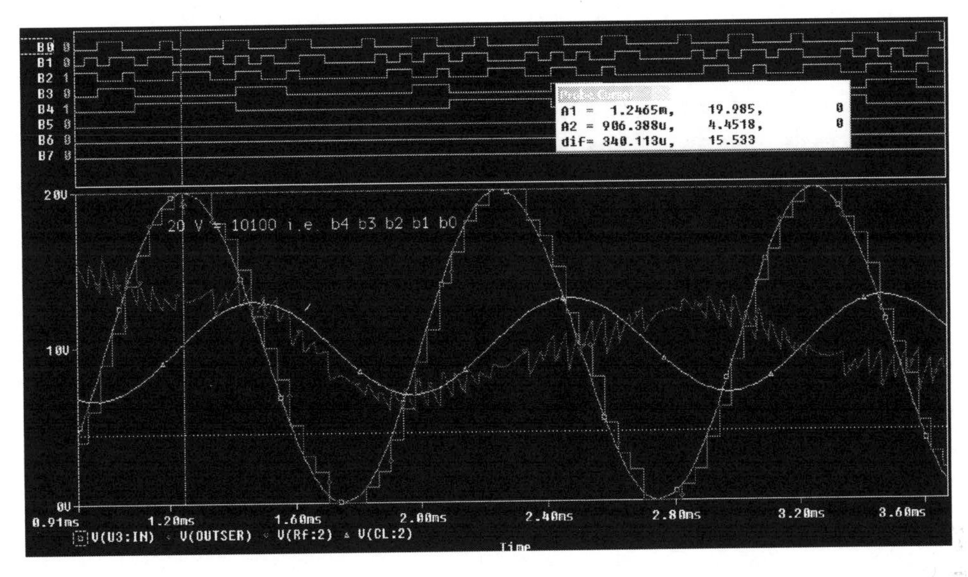

FIGURE 3.16: 20 V input voltage.

The signal in Fig. 3.16 shows the input 10 V offset raising the 10 V AC input. The binary signal is read vertically downward using the cursor located at the maximum input signal of 20 V.

3.5.1 DAC Resolution

The DAC output is calculated as

$$V_{\text{out}} = V_{\text{ref}}\left[\frac{D7}{2}+\frac{D6}{4}+\frac{D5}{8}+\frac{D4}{16}+\frac{D3}{32}+\frac{D2}{64}+\frac{D1}{128}+\frac{D0}{256}\right] V. \tag{3.25}$$

The output voltage for a binary DAC output 00001001 and 256 V reference voltage is

$$V_{\text{out}} = V_{\text{ref}}\left[\frac{0}{2}+\frac{0}{4}+\frac{0}{8}+\frac{0}{16}+\frac{1}{32}+\frac{0}{64}+\frac{0}{128}+\frac{1}{256}\right] = 256(1/32+1/256) = 9\ V. \tag{3.26}$$

The **DigClock** conversion clock has on/off times equal to 20 μs. Set the transient parameters as follows: **Output File Options/Print values in the output file** = 1ms, **Run to time** = 20ms, **Start saving data after** = 2ms. Making the **Output File Options/Print values in the output file** larger than the 20 ns default value will speed up the simulation time in most instances. However, it must be set to a value smaller than the **Run to time**. If all eight bits are used, i.e., full scale, then we will have 256 levels available ($2^8 = 256$). Table 3.1 shows the relationships between the ADC resolution and the voltage reference Vref.

TABLE 3.1: The ADC and Reference Relationship

VIN (V)	VREF (V)	NO. OF BITS	NO. OF LEVELS	RESOLUTION (V)
9	16	8	$2^8 = 256$	0.0625
9	32	7	$2^7 = 128$	0.125 V
9	64	6	$2^6 = 64$	0.25
9	128	5	$2^5 = 32$	0.5
9	256	4	$2^4 = 16$	1

The input signal **VSIN** generator parameters are **VOFF** = 5 V, **VAMPL** = 4 V, and **FREQUENCY** = 1 kHz. To measure the quantization noise, we must attenuate the 1 kHz input signal using a band-stop notch filter.

3.5.2 Band-Stop Filter

The band-stop (notch) filter is implemented using an **ABM** part called **BANDREJ**. The filter is characterized by six parameters: four cut-off frequencies and two attenuation values—**RIPPLE** defines the maximum allowable ripple in the passband and **STOP** is the minimum attenuation in the stopband. Set the Analysis tab to Analysis type: AC Sweep/Noise, AC Sweep Type to **Linear/Logarithmic**, **Start Frequency** = 100, **End Frequency** = 10 k, **Total Points/Decade** = 10001. Enter the lowest frequency F0 to highest frequency F3 for the band-stop filter parameters as shown in Fig. 3.17. A value for either of the passband frequencies, say f_{p1}(or stop frequencies), is calculated by assuming a value for one of the passband frequencies and knowing the center frequency to yield for f_{p2}

$$f_0 = \sqrt{(f_{s1} f_{s2})} = \sqrt{(f_{p1} f_{p2})} \Rightarrow f_{p2} = f_0^2 / f_{p1}. \tag{3.27}$$

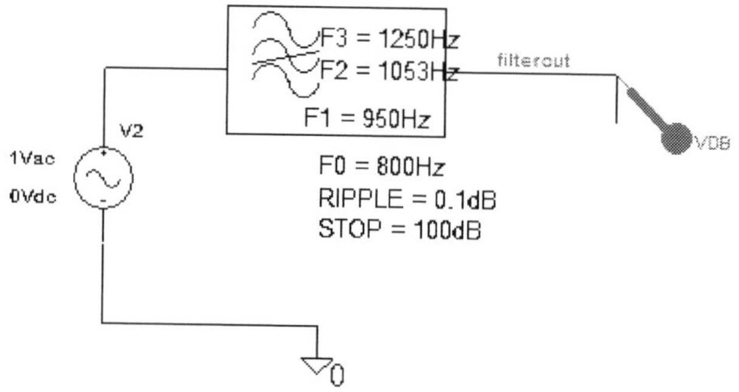

FIGURE 3.17: The band-stop part called **BANDREJ**.

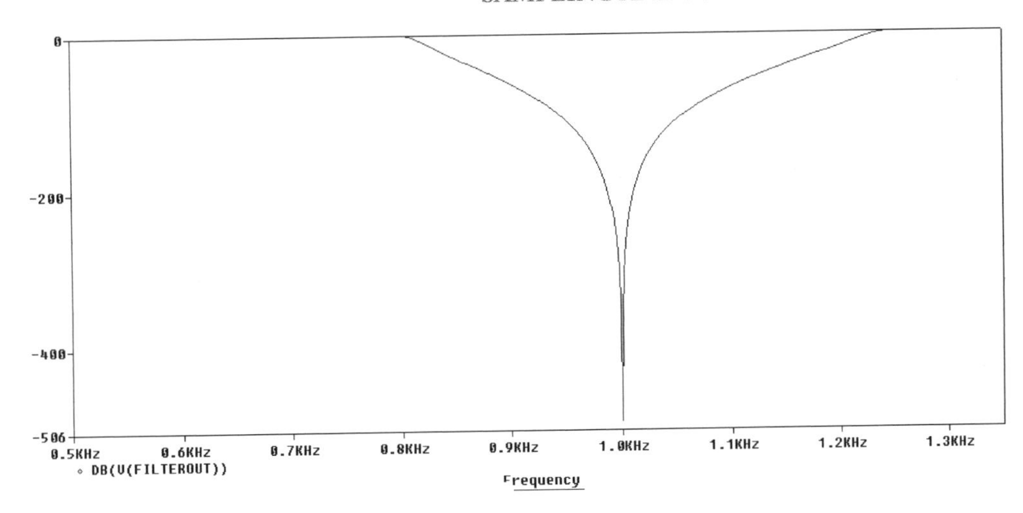

FIGURE 3.18: The band-stop amplitude response.

The band-stop filter response is shown in Fig. 3.18.

We are now in a position to investigate the relationship between the SQNR and the number of bits used. Delete all traces in the **PROBE** screen and insert a trace, which is the RMS of the output signal from the band-stop filter. Measure the noise RMS value and repeat for a smaller number of bits, by deleting the lines b0 to b7 from the input of the DAC and in turn measuring the SQNR. Alternatively, you may place switches on each bit line, and then open each one in succession.

3.6 PULSE CODE MODULATION

It is not economical to send signals in parallel form over a telephone cable network, so the output from the analog to digital converter (ADC) is multiplexed into a serial bit stream. The multiplexing and demultiplexing process in Fig. 3.19 is an important part of the 8-bit 30-channel TDM PCM system and requires 8-bit parallel to serial conversion. There are many multiplexing IC devices but here we use the 74151 IC where the IC pins S0, S1, and S2, control the sequence of conversion. The **STIM4** generator part has four parallel output lines, of which one is unused but is terminated with a line segment name called unused. Command line 1 starts the sequence as 0ms and the parallel bit pattern is 0000. The four parallel signals repeat a number of times using a repeat function on command line 2 and the sequence is terminated on command line 4 with **endrepeat**. Command line 3 shows the pattern being incremented by 1 every 125 μs.

The generator output is terminated with a **BUS** line using the **Place Bus** icon (Short-cut key B). **Dlclick** the bus and type in s[3-0] where each line connected to the bus is named s0,

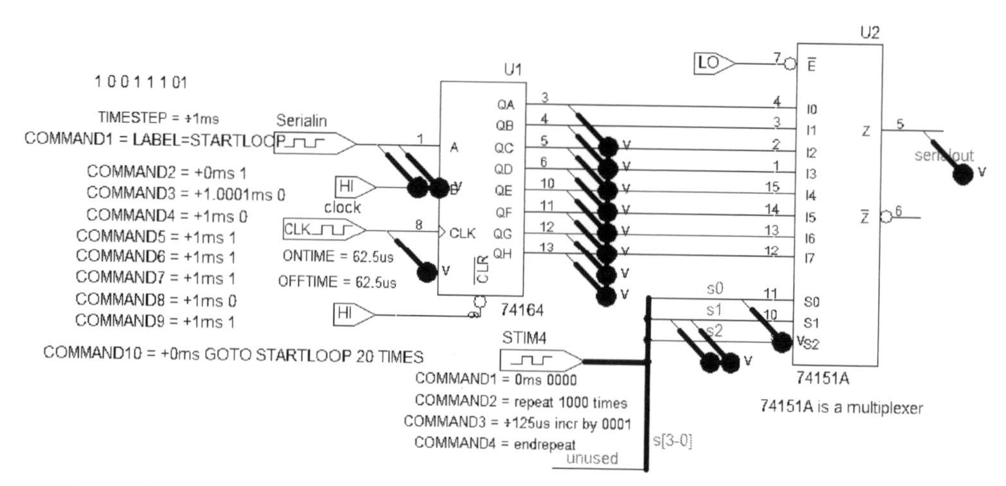

FIGURE 3.19: Multiplexing and demultiplexing.

s1, s2, and s3. Placing a marker on a bus produces a **PROBE** display in hexadecimal format. The clock frequency is twice the speed of the **STIM4** part increment (i.e. half the time period). In an actual system, the final output s3 could synchronize the system. The serial data is applied using a **STIM1** part.

Set the Analysis tab to Analysis type: **Time Domain** (Transient), **Run to time** = 10ms, and **Maximum step size** = (left blank), Press F11 to simulate. The waveforms for the **mux–demux** devices are shown in Fig. 3.20.

3.6.1 Universal Shift Register

74194 LS is a 4-bit bidirectional universal shift register where all data and mode control inputs are edge-triggered and respond to LOW to HIGH clock transitions only. The mode control

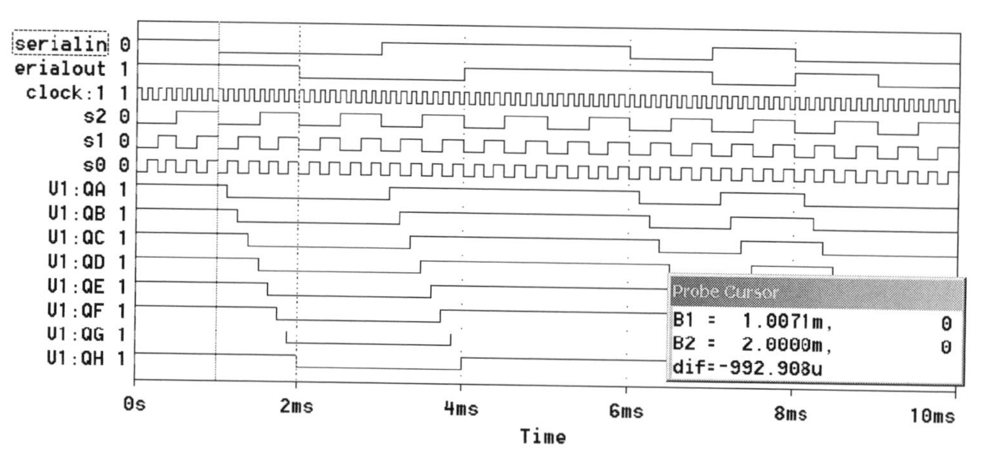

FIGURE 3.20: Serial to parallel and parallel to serial conversion.

TABLE 3.2: 74194 Universal Shift Register

OPERATING MODE	CLR (1)	S1 (10)	S0 (9)	SR (2)	SL (7)	PN (3–6)	QA (15)	QB (14)	QC (13)	QD (12)
Reset	L	X	X	X	X	X	L	L	L	L
Hold	H	L	L	X	X	X	Q0	Q1	Q2	Q3
Shift left	H	H	L	X	L	X	Q1	Q2	Q3	L
	H	H	L	X	H	X	Q1	Q2	Q3	H
Shift right	H	L	H	L	X	X	L	Q0	Q1	Q2
	H	L	H	H	X	X	H	Q0	Q1	Q2
Parallel load	H	H	H	X	X	Pn	P0	P1	P2	P3

and selected data inputs must be stable one setup time prior to the positive transition of the clock pulse. The four parallel data inputs, P0, P1, P2, P3, are D-type inputs (also called A, B, C, and D). When both S0 and S1 are HIGH, signals presented to P0, P1, P2, and P3 inputs are transferred to Q0, Q1, Q2, and Q3 outputs, respectively, following the next LOW to HIGH transition of the clock. This is a parallel-to-parallel conversion. The mode control inputs, S0 and S1, determine the synchronous operation of the device. Table 3.2 shows how data is shifted from left to right (Q0 → Q1), or right to left (Q3 → Q2). Parallel data is entered, by loading all four bits of the register simultaneously.

When both S0 and S1 are LOW, the existing data is retained in a "do-nothing" mode without restricting the HIGH to LOW clock transition. D-type serial data inputs (SR, SL) allow multistage shift right or shift left data transfers without interfering with parallel load operation, and will appear on the QA and QC outputs. The asynchronous clear (CLR), when LOW, overrides all other input conditions and forces the Q outputs LOW.

In the table, L = low level, H = high level, X = do not care, l = LOW voltage level one setup time prior to the LOW to HIGH clock transition; h = high voltage level one setup time prior to the LOW to HIGH clock transition; P(n) and Q(n), where n indicates the state of the input or output one setup time prior to the LOW to HIGH clock transition.

3.6.2 74194 Universal Shift Register

Draw the schematic in Fig. 3.21. What is the output for the parallel input 0110? Set the Analysis tab to Analysis type: **Time Domain** (Transient), **Run to time** = 10ms, and **Maximum step size** = (left blank), press F11 to simulate.

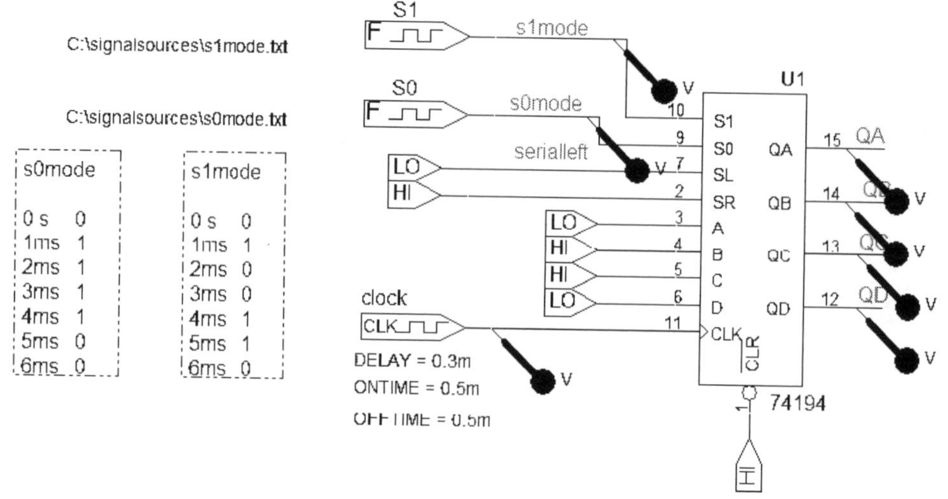

FIGURE 3.21: Universal shift register.

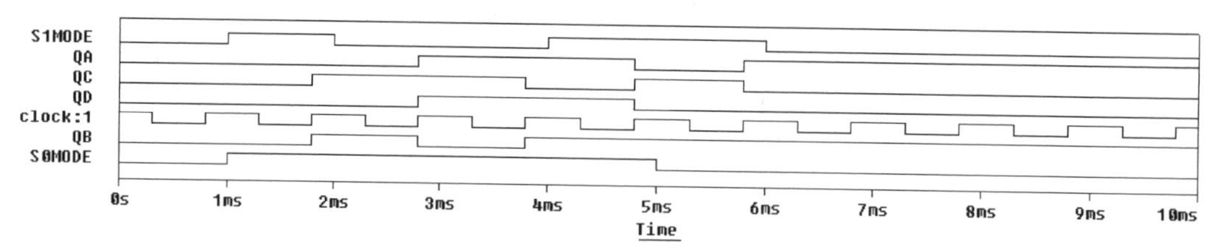

FIGURE 3.22: Universal shift register signals.

3.7 SINGLE-CHANNEL 4-BIT PCM TRANSMITTER

The simple PCM system in Fig. 3.23 uses only four of the 8 bits available, but it may be extended to an 8-bit capability using additional IC circuitry (see exercise at the end of the chapter). Here, we apply a 1 kHz sinusoidal signal and sample it at 8000 Hz, which is the rate used in the public switched telephone network (PSTN) and converts the ADC parallel output to a serial bit stream. To test that the circuit is working correctly, we need to reverse the procedure and attach a serial to parallel converter, a latch for temporary store, and a DAC.

The DAC output is then low-pass filtered. The registers must be initialized to a certain state. Select the **Simulation Setting** menu and then select the **Options** tab. In the **Category** box select **Gate-Level Simulation.** This will then show the **Timing Mode** - select **Typical**. In the **Timing Mode** change **Digital Setup** from **Typical** to **Maximum**. The **Initialize all flip flops** is set to **All 1**. Failing to do this will result in the flip-flop being in an undetermined state, i.e., **All X**, and shows up in the plot as two red lines. Set the Analysis tab to Analysis type: **Time Domain** (Transient), **Run to time** = 2ms, and **Maximum step size** = (left blank), Press F11 to simulate.

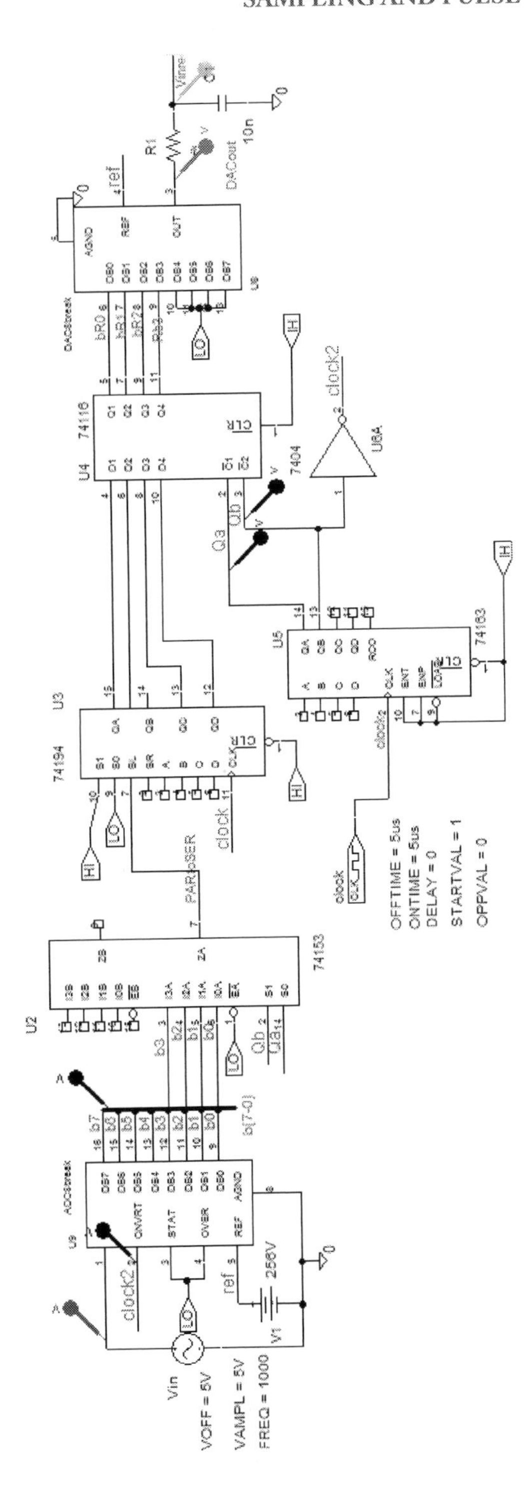

FIGURE 3.23: Single-channel 4-bit PCM transmitter and receiver.

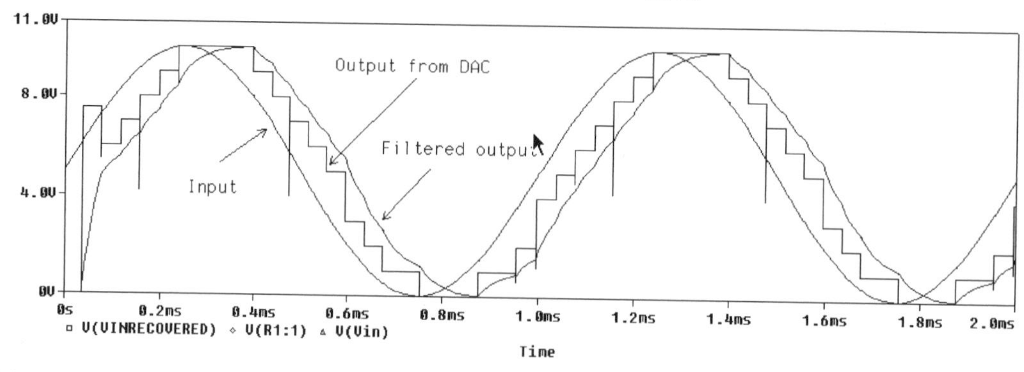

FIGURE 3.24: Input and recovered signals.

The recovered 1 kHz sinusoid in Fig. 3.24 has distortion that is displayed using the **FFT** function.

The timing waveforms should be investigated.

3.8 TIME-DIVISION MULTIPLEXING AND DEMULTIPLEXING

In the European E1 PCM telephone systems, 30 voice signals are sampled, coded, and multiplexed, to produce a 2.048 Mb/s digital signal that is transmitted over a high-bandwidth line, coaxial cable, microwave line-of-sight link, optical fiber, etc. Sample binary values are multiplexed using byte interleaving so every 125 μs a framing bit is followed by a coded sample for each of the 30 voice signals. Other housekeeping bits are allocated for different purposes. Also super frames take care of the signaling requirements by 16 frames with the center time slot used for two channel signaling of 4 bits each. The transmission rate is 257 bits every 125 μs (2.048 Mb/s $= 8000 \times 8 \times 32$). The U.S. and Japanese 24-channel time-division multiplexed (TDM) PCM system generates a composite bit stream for a total of 24 digitized voice channels and results in a frame structure consisting of 193 bits in each 125 μs time interval. Secondary multiplexed systems combine primary multiplexing digital streams into a higher rate output bit stream called the plesiochronous digital hierarchical structure. A TDM transmitter and receiver, using **ABM** library parts, are shown in Fig. 3.25. Set the Analysis tab to Analysis type: **Time Domain** (Transient), **Run to time** = 4ms, and **Maximum step size** = (left blank), Press F11 to simulate.

Fig. 3.26 shows the TDM and recovered signals.

3.8.1 Time-division Multiplexing of Two PAM Signals

The 4066 IC in Fig. 3.27 multiplexes two signals. One of the electronic switches may be used (see exercise 12 at the end of the chapter) to invert the clock so that the second signal is sampled with a delay of one clock period.

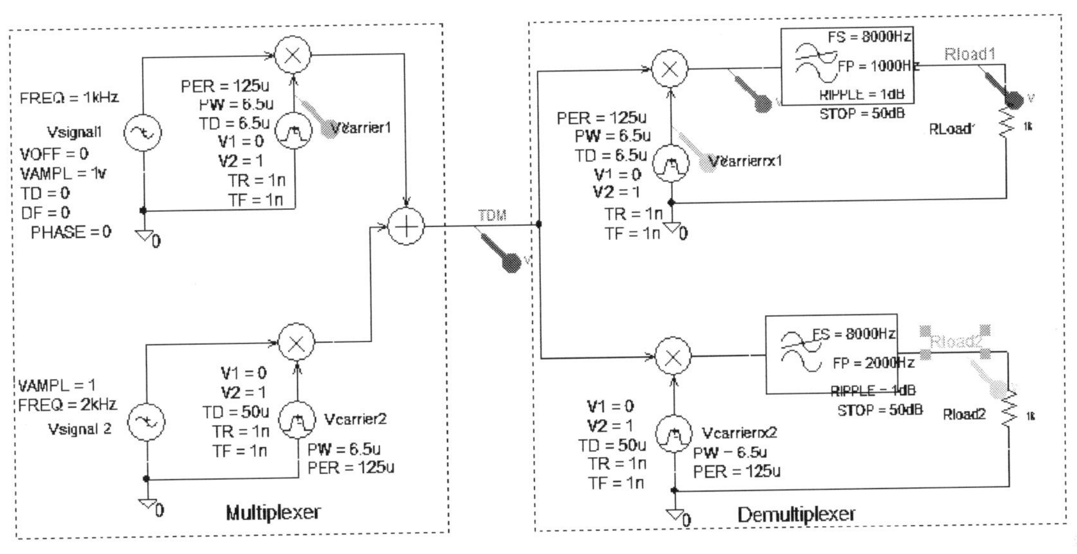

FIGURE 3.25: Time-division multiplex systems.

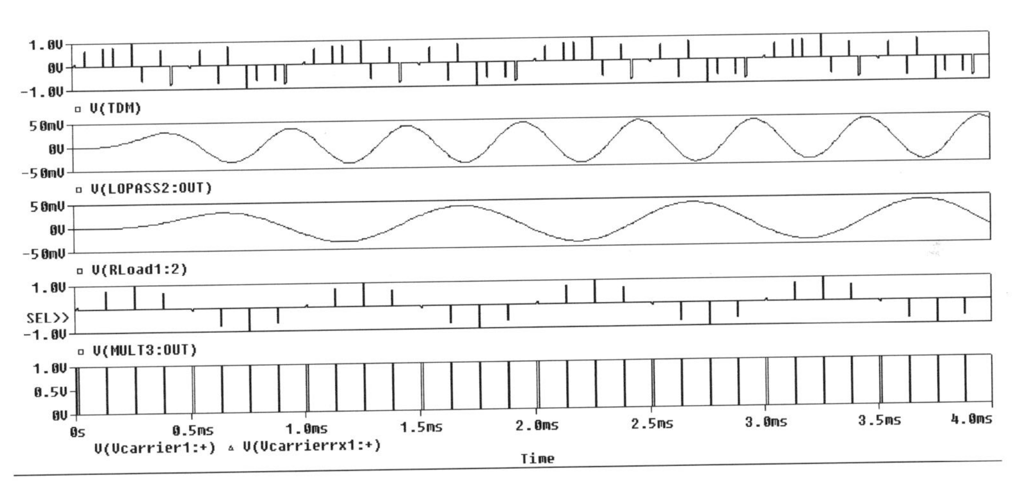

FIGURE 3.26: TDM waveforms.

You may get a "too-large fatal error" using the evaluation PSpice, so use the alternative multiplexing circuit in Fig. 3.29, or substitute **ABM** switches instead of the 4016. Be careful about using analog parts with digital circuitry. PSpice inserts analog to digital converters automatically when analog and digital parts are connected together. This increases the number of components and hence reduces the size of the circuit you may use in the evaluation version. For example, use a **DigClock** generator instead of the **VPULSE** generator. Doing this means that you may place extra components before the evaluation limitations are reached. Set the

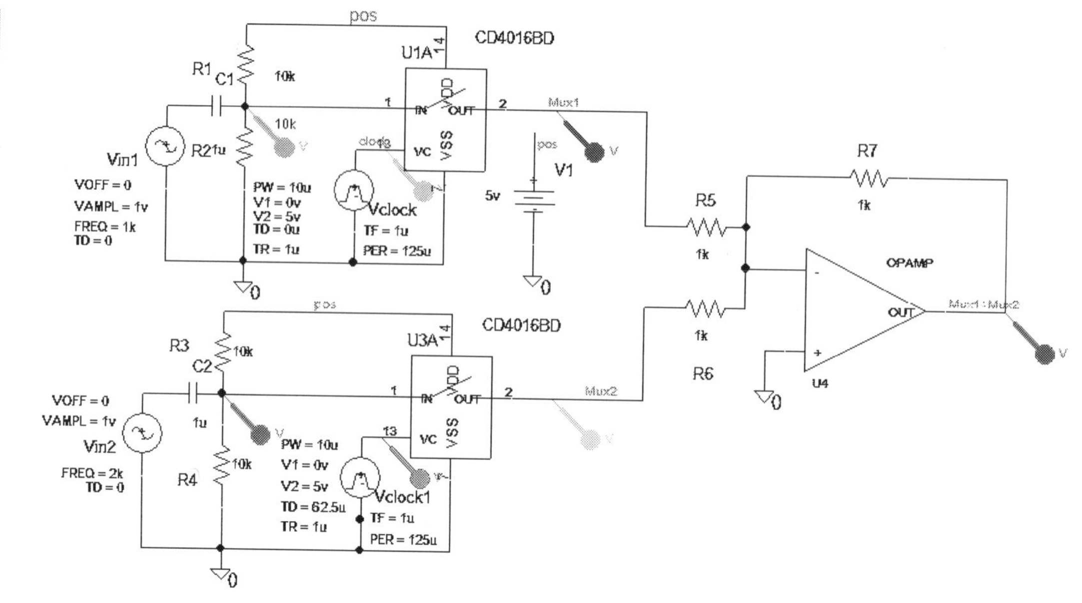

FIGURE 3.27: Multiplexing two signals.

Analysis tab to Analysis type: **Time Domain** (Transient), **Run to time** = 2ms, and **Maximum step size** = (1us), press F11 to simulate.

The TDM signals are shown in Fig. 3.28.

3.9 LINEAR DELTA MODULATION

The delta modulation in Fig. 3.30 is a form of differential PCM produced by comparing the input signal with an approximation of this signal in a comparator. This produces an error signal that is sampled and quantized to a positive or negative signal. This signal generates an approximate signal using a decoder and is transmitted to a remote decoder to produce an identical signal if no errors are present in the receiver. After each sample, the approximate signal increments or decrements by the step size. A fixed-step size may cause slope overload but is corrected using adaptive delta modulation.

The comparator output is a delta-modulated carrier signal connected to a Schmitt trigger, and the flip-flop output is integrated and is one of the inputs to the comparator. Set the Analysis tab to Analysis type: **Time Domain** (Transient), **Run to time** = 2ms, and **Maximum step size** = (left blank), Press F11 to simulate. After simulation, you may notice Simulation Message warnings popping up. Select **Yes** and one of the messages in Fig. 3.31 appears.

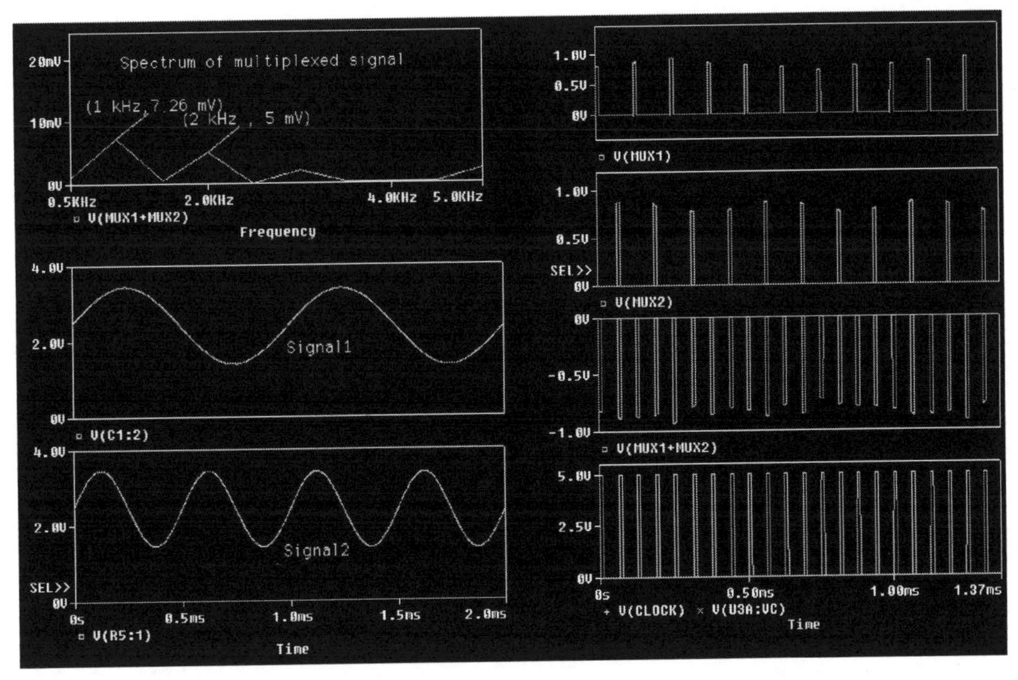

FIGURE 3.28: TDM over an extended period.

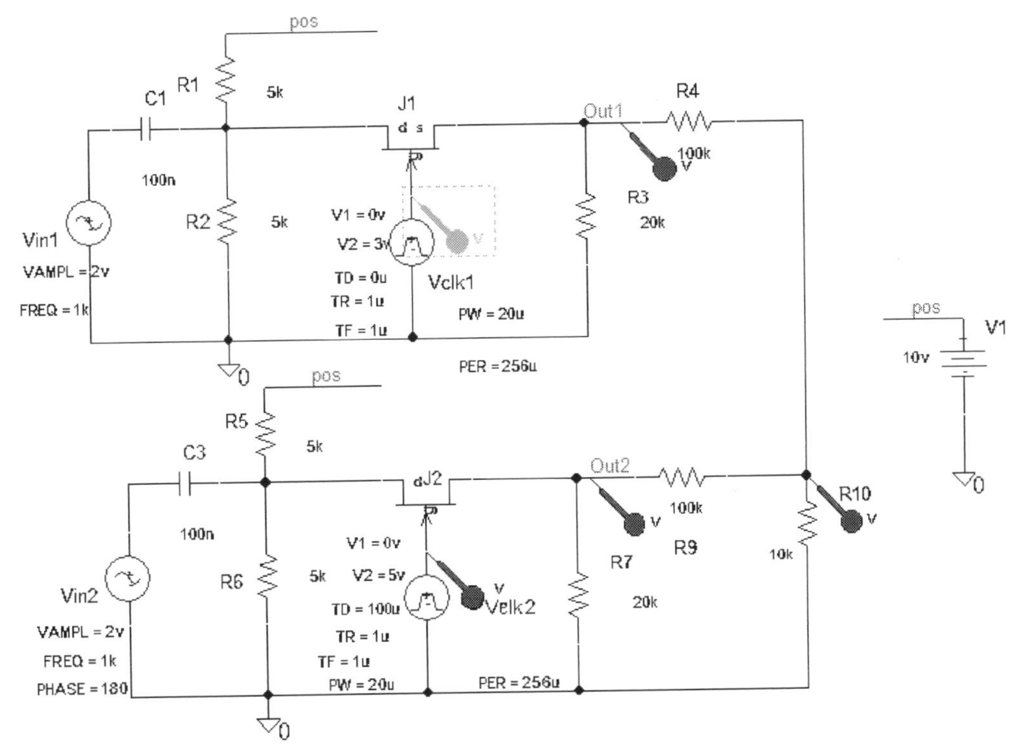

FIGURE 3.29: TDM circuit using FET switches.

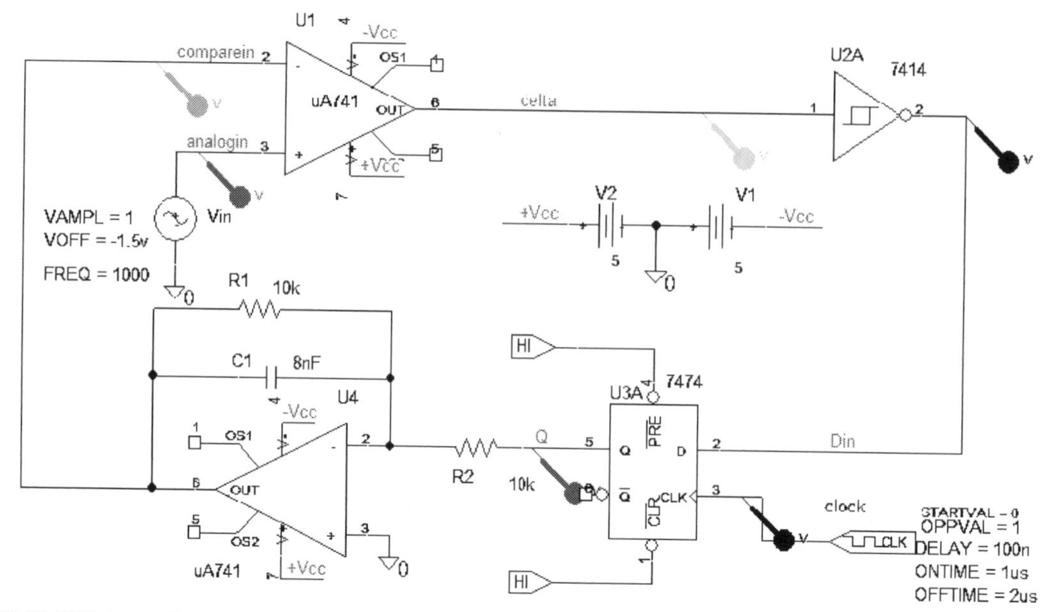

FIGURE 3.30: Integrating delta modulator.

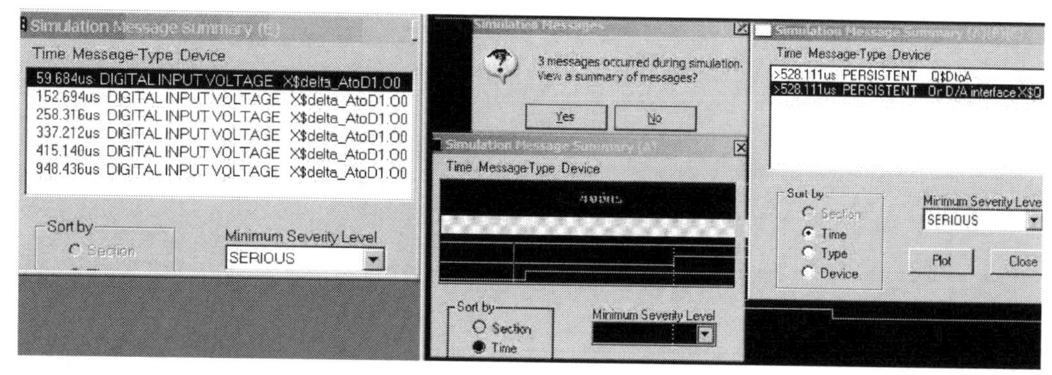

FIGURE 3.31: Specific warning messages.

These are serious warnings and should not be ignored if the circuit is to be realized in hardware. The warnings are not always obvious but in this example when you select **Plot**, you see the message "7.000 is beyond the ranges defined in model." The input signal to the Schmitt-trigger inverter is too large in this case and should be limited. Warnings will also flag race hazards and other problems associated with digital circuits. We may limit the signal amplitude using a Zener regulator placed in front of the Schmitt trigger as shown in the transmitter–receiver shown in Fig. 3.33. To see other warnings, set the clock ONTIME = OFFTIME = 1 μs and simulate. Observe the integrating delta modulator signals in Fig. 3.32.

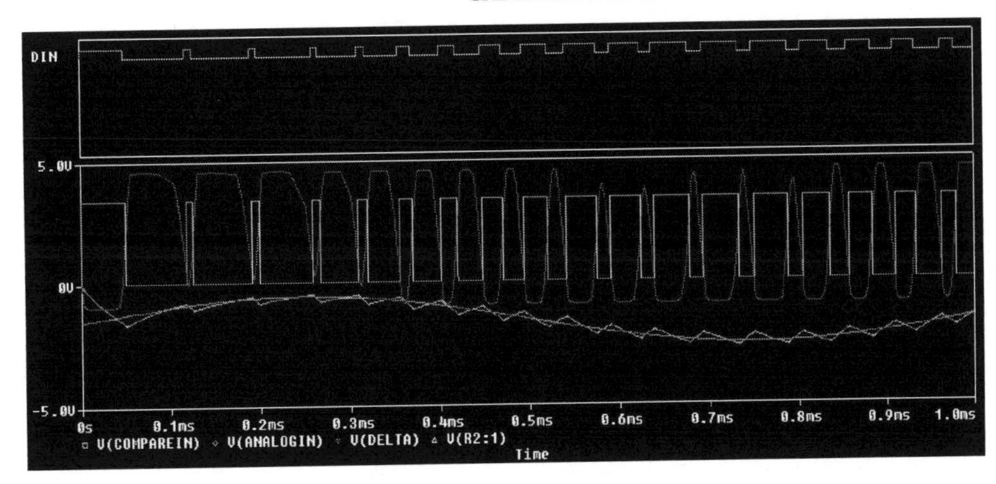

FIGURE 3.32: Integrating delta modulator waveforms.

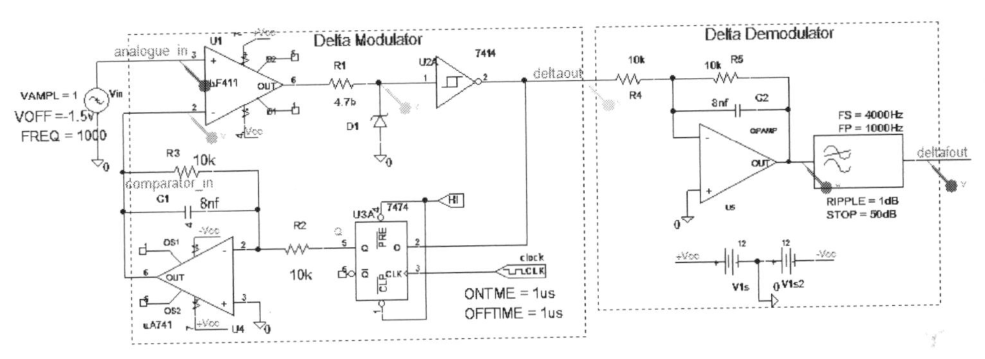

FIGURE 3.33: Delta demodulator.

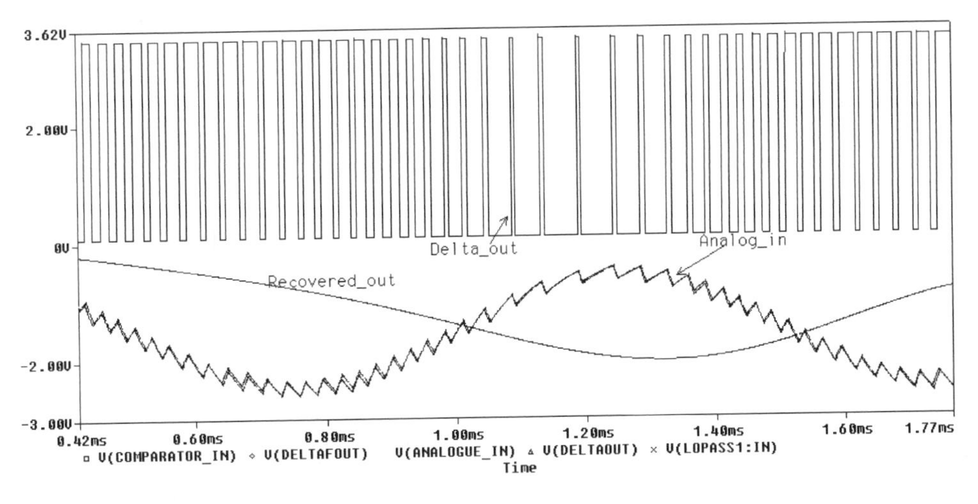

FIGURE 3.34: Delta demodulator waveforms.

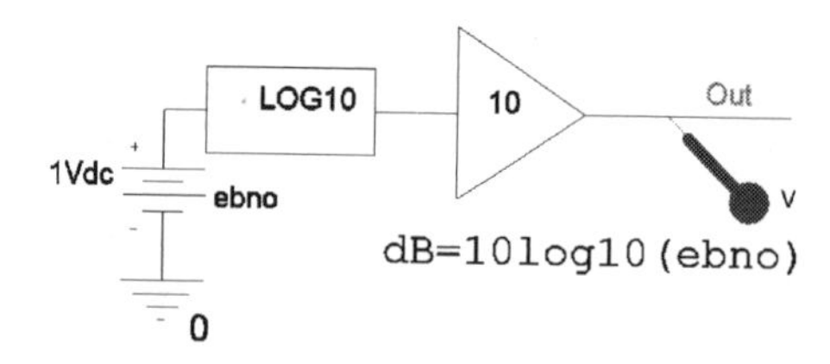

FIGURE 3.35: **ABM Log10** part.

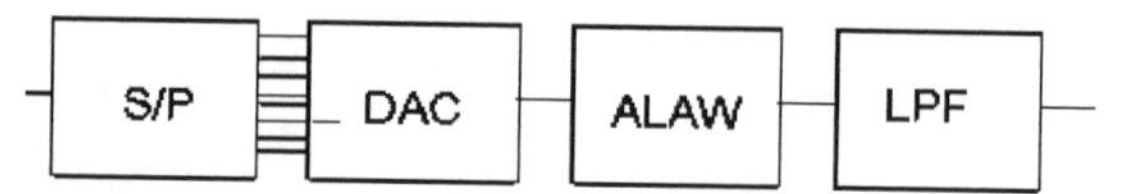

FIGURE 3.36: PCM receiver.

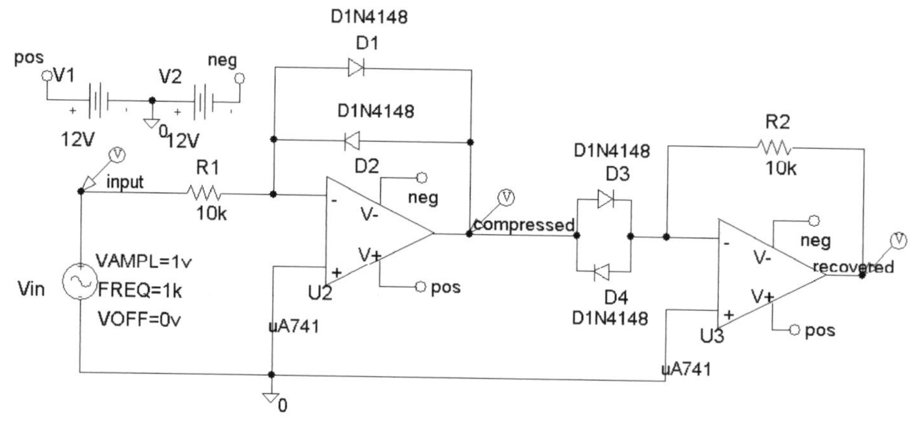

FIGURE 3.37: Compressor and expander.

Slope overload may occur in a fixed-step size delta modulator but is overcome using a variable step size.

3.9.1 Delta Demodulation

Fig. 3.33 shows that a delta demodulator is accomplished using an identical integrator and an **ABM** LPF (**LOPASS** part). To get this to work in the evaluation version requires using the ideal **OPAMP** part.

The demodulator output is shown in Fig. 3.34.

3.10 EXERCISES

1. To achieve better sine extraction, redesign the filter in Fig. 3.15 using a Chebychev design (better roll-off rate than the Butterworth configuration).

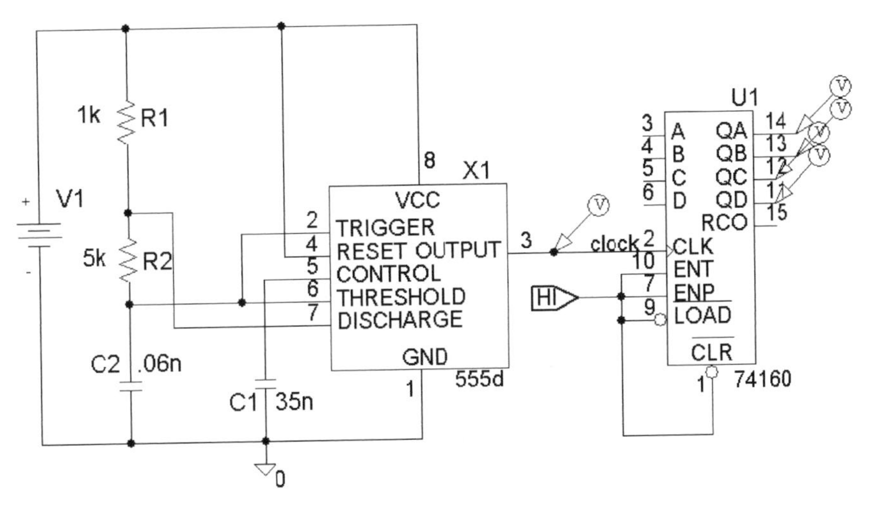

FIGURE 3.38: 4-bit counter.

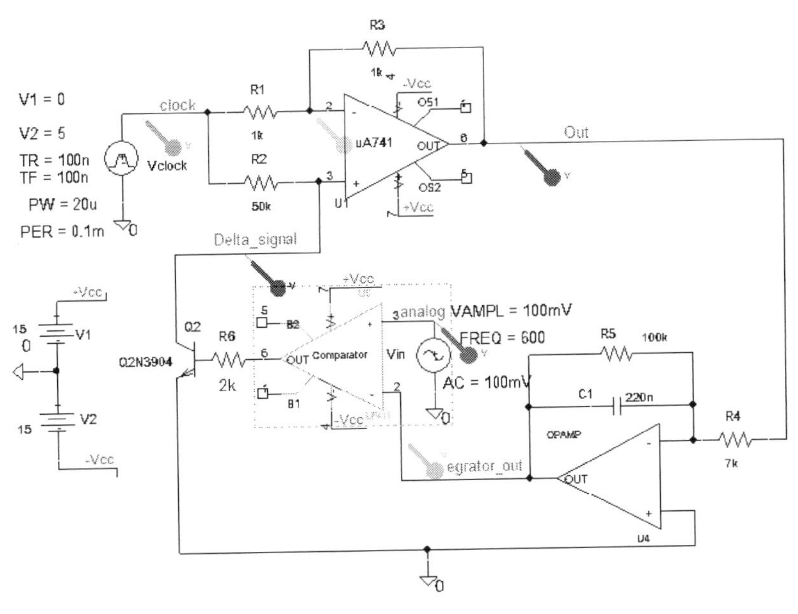

FIGURE 3.39: Integrating delta modulator.

2. Investigate analog compression using the log10 part as shown in Fig. 3.35.
3. Replace the **ABM** band-stop filter in the single channel PCM system with an actual circuit notch filter. Investigate how you would replace **STIM4** in Fig. 3.19 with actual circuitry (hint: 74163 + 555). If you get stuck with this exercise see exercise (6).
4. Draw a hierarchical schematic showing the elements of the PCM receiver in Fig. 3.36.
5. Investigate the compressor and expander schematic shown in Fig. 3.37.

6. The schematic in Fig. 3.38 is a 4-bit counter with a 555 IC reference clock. Investigate this circuit and use it to replace the **STIM4** part in the serial to parallel conversion discussed previously. How could this schematic be modified or extended to produce an 8-bit counter?
7. The output from the comparator in Fig. 3.39 is a delta-modulated carrier signal and is produced by comparing the input signal with the output signal from the staircase generator. A ramp approximation is produced by delivering a constant current to the capacitor throughout the sampling period, the current polarity being determined at the sampling instant. A staircase approximation to the input signal is produced by delivering (or extracting), a unit of charge at each sampling instant.
   A train of pulses using a **VPULSE** generator is connected to the circuit with parameters as shown. The clock period is approximately a tenth of the input analog signal period. The amplitude of the 600 Hz sinusoidal analog signal is 0.1 V. Set the transient parameters as follows: **Run to time** = 1.5ms, and **Maximum step size** = 1 μm. From the **PROBE** screen, select the **Trace** icon and add traces as observed in Fig. 3.40.
8. Investigate the universal IC 74194 as a 4-bit P-to-S, S-to-P, or P-to-P configuration. Table 3.3 shows the signals applied to the mode setting pins.
9. Modify the previous schematic and replace the 1 kHz signal with two multiplexed speech signals.
10. A two-channel TDM circuit is shown in Fig. 3.41. Extend the multiplexing to four channels and add a receiver.

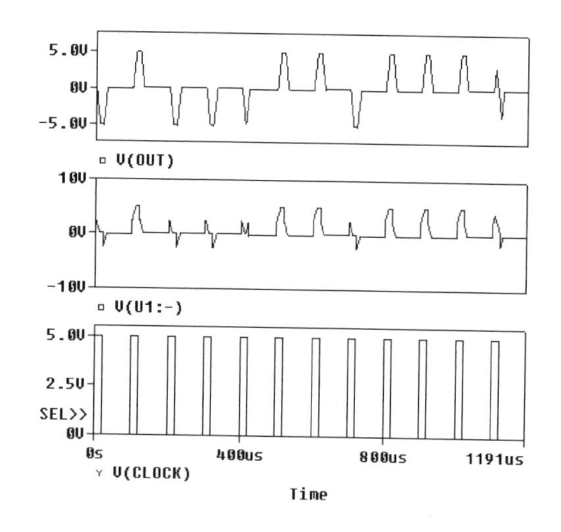

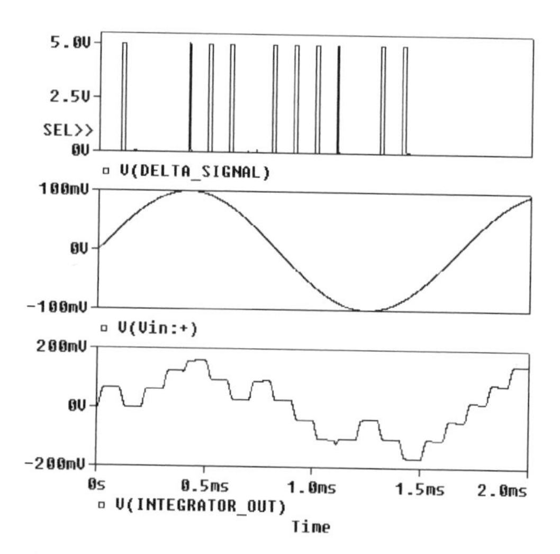

FIGURE 3.40: Delta modulator time diagrams.

TABLE 3.3:

MODE	SO	S1
Hold	0	0
Shift left	0	1
Shift right	1	0
Parallel load	1	1

The bus line is normally reserved for digital signals but works for analog signals as well.

11. Investigate the 8-bit PCM system in Fig. 3.42. One of my students from India, SingaravelanVaithiyanathan modified my basic system to create this 8-bit system. It is not a true 8-bit system but a good attempt nevertheless. Redraw the schematic but use hierarchical blocks.
    Because of the complexity and number of parts, this schematic will display "FATAL PSpiceAD 11:08AM Fatal Digital Simulator Error 4." Select **File** from PSpiceAD

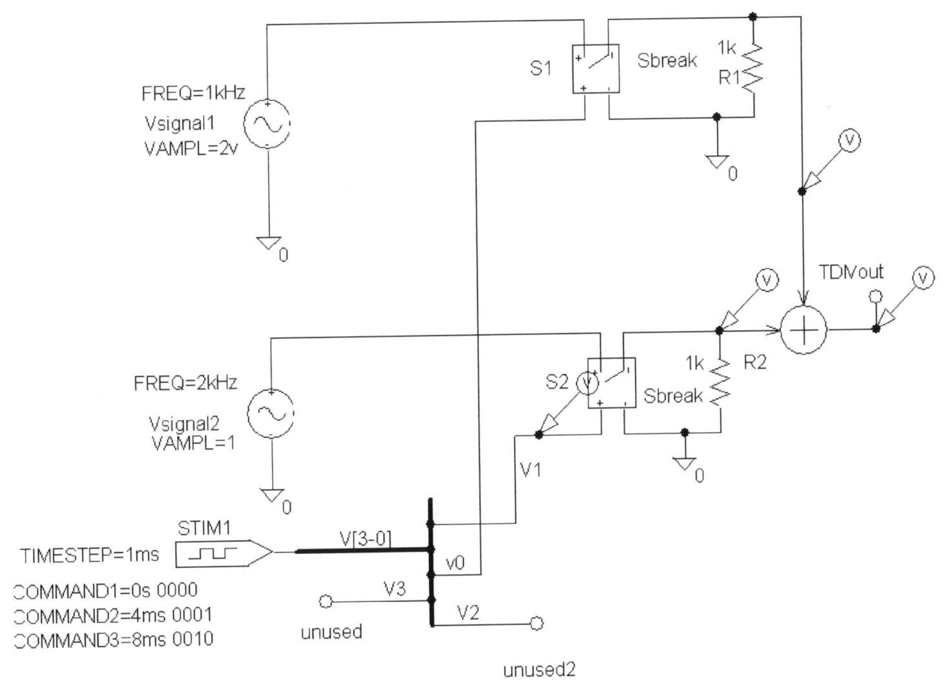

FIGURE 3.41: TDM: note the use of the bus line.

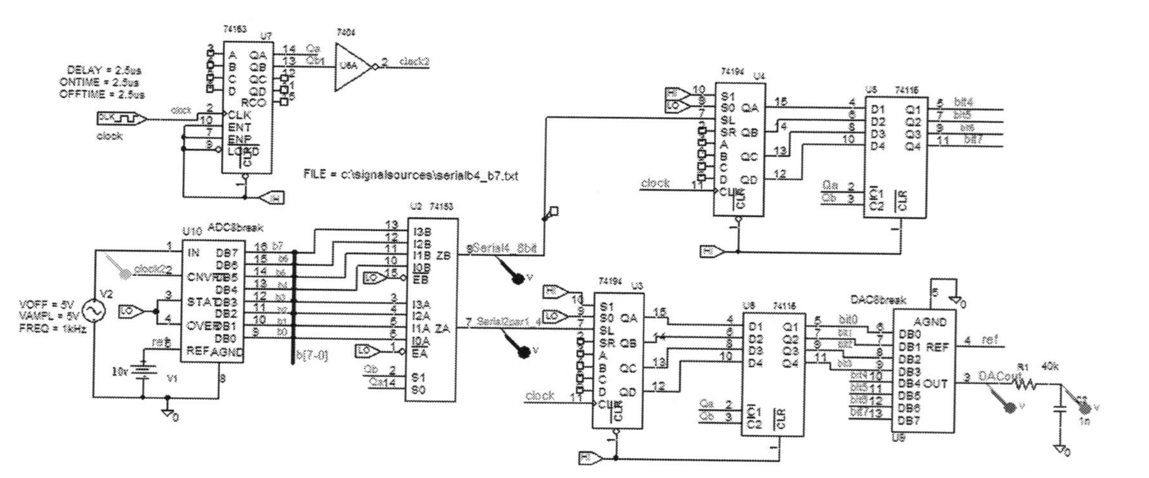

FIGURE 3.42: 8-bit PCM system.

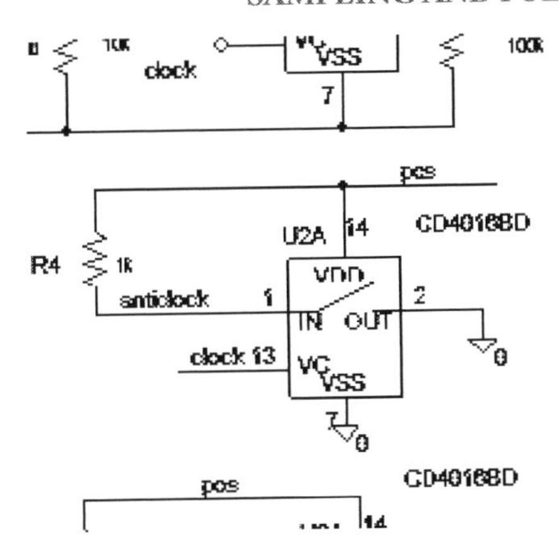

FIGURE 3.43: Obtaining an inverted clock.

and Run **PROBE** to show the limited response. A **VECTOR** part records the output and a **FileStim** part applies the recorded signal to the second part of the system.

12. Investigate the method of inverting a clock shown in Fig. 3.43.

CHAPTER 4

Passband Transmission Techniques

4.1 BASEBAND TO PASSBAND

Baseband communication techniques for encoding local area and telephone network data signals were examined in previous chapters, but here we investigate passband modulation. A baseband signal cannot be transmitted over free space, or a network with a poor low-frequency response, and so the data must be modulated thus creating a passband signal. Fig. 4.1 shows an encoded baseband signal $x(n)$ converted to a passband signal $s(n)$ where the data modulates a high-frequency carrier.

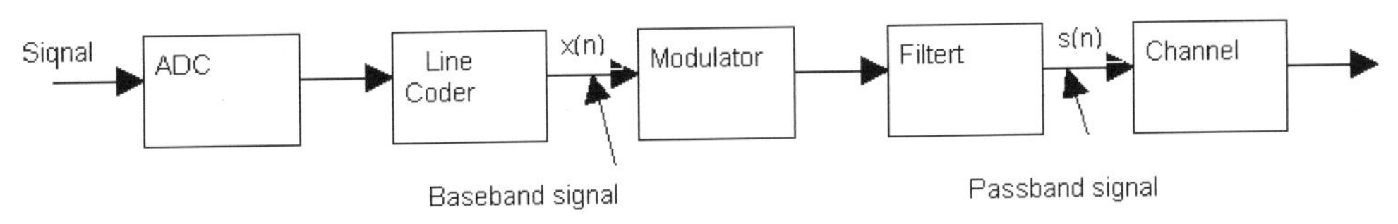

FIGURE 4.1: Baseband to passband conversion.

Modulating a high-frequency carrier, with a baseband signal, produces a signal with upper and lower sidebands. For example, in Fig. 4.3 the passband spectrum has a bandwidth that is twice that of the baseband signal. Some of the passband systems investigated in this chapter are a bit outdated but are used in sub systems in more complex transmission systems such as frequency-shift keying (FSK) in spread spectrum systems (chapter 7).

4.2 AMPLITUDE SHIFT KEYING

Amplitude shift keying (ASK) amplitude modulates a carrier f_c using binary data, $d(t) = \pm 1$ V. The instantaneous carrier amplitude of an ASK signal is

$$\text{ASK}(t) = \{1 + d(t)\}\frac{A}{2}\sin(2\pi f_c t)\ V. \tag{4.1}$$

Equation (4.1) is zero for $d(t) = -1$, and $A\sin(2\pi f_c t)$ for $d(t) = 1$. Fig. 4.2 shows the **Sbreak** switch activated by the baseband data **VSIN** part. The Vdata then modulates a 100 kHz carrier to produce the ASK signal. Identifying the different signals displayed after simulation is

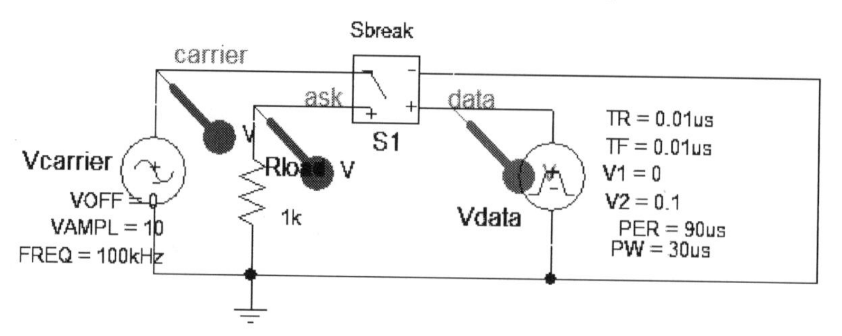

FIGURE 4.2: Amplitude shift keying.

made easier by naming each wire segment. Select each wire segment and, using the **Net Alias** icon, enter the wire segment names carrier, **ask**, and **data**.

The on/off resistances simulate real-world switch resistance values having a resistance of 1 MΩ when opened, and 1 Ω when closed. The data is simulated using a **VPULSE** part labeled Vdata as shown. Modulation depth (or index) variation is achieved by changing the data amplitude V2 from 0.1 V to 1 V. The largest V2 value produces a special form of ASK called on–off keying (OOK). Set **Output File Options/Print values in the output file** to 0.1 μs, **Run to time** to 300ms, and **Maximum step size** to 10 μs. Press **F11** to simulate and display the FSK signal in the time and frequency domains in Fig. 4.3. The ASK spectrum shows the

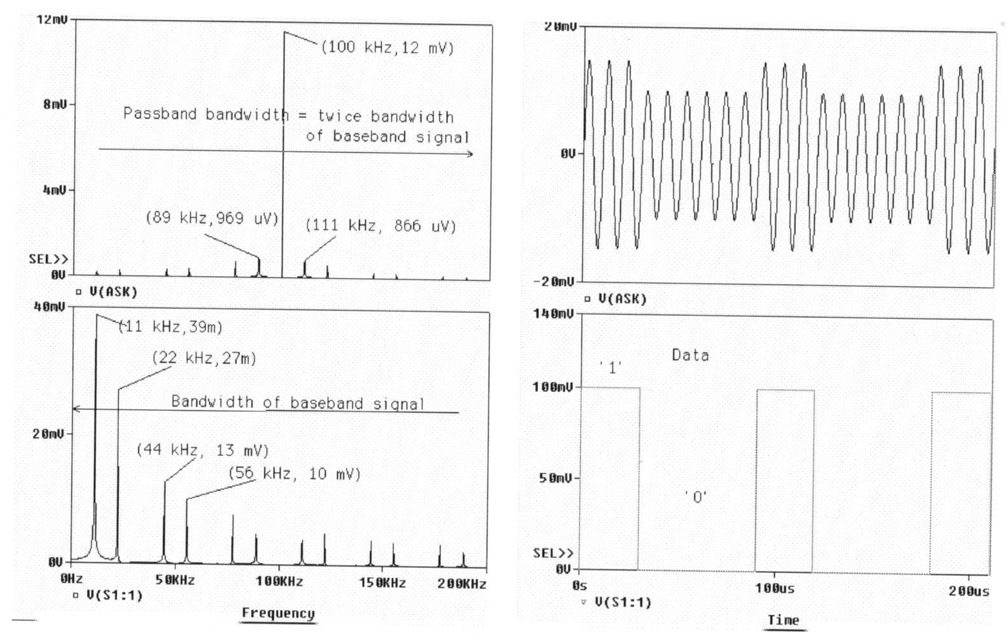

FIGURE 4.3: ASK time signals with V2 = 1 V.

baseband data sinc-shaped spectrum shifted to a higher frequency, thus creating a passband signal.

4.3 FREQUENCY SHIFT KEYING

A form of frequency shift keying (FSK) is used in frequency-hopping spread spectrum systems (the DECT domestic phone system), and certain satellite systems. FSK is similar to frequency modulation in that the data modulates a carrier producing two carrier frequencies for each of the signal conditions. One frequency represents 1 (or mark) and another frequency represents 0 (or space), in the baseband data signal. The instantaneous value of the FSK carrier signal is expressed as

$$\begin{aligned} V_{\mathrm{FSK}}(t) &= V_c \sin(2\pi f_m t) \quad \text{for} \quad \text{"1"} \\ V_{\mathrm{FSK}}(t) &= V_c \sin(2\pi f_s t) \quad \text{for} \quad \text{"0"}. \end{aligned} \tag{4.2}$$

In general,

$$v_{\mathrm{FSK}}(t) = V_c \sin 2\pi (f_{c1} + s(t)\Delta f_{c1}) t \text{ V}. \tag{4.3}$$

Fig. 4.4 shows how an FSK signal is produced using the **EVALUE** part as a voltage-controlled oscillator (VCO).

The carrier signal is deviated 1 kHz above and below the 3 kHz carrier signal, by the data signal generated using a **VPULSE** generator part. Select the V(%IN+, %IN−) part of the **EVALUE** component and substitute in the **EXPR** box: SIN(2*pi*3k*TIME*V(%IN+, %IN−)). The sine angle, and hence the carrier frequency, is a function of the input controlling voltage V(%IN+, %IN−). The 3 kHz carrier is deviated by the Vmod generator controlling voltage parameters. The magnitudes of the modulating voltage, Vmod, are 2 kHz/3 kHz = 0.666 V and 4 kHz/3 kHz = 1.333 V. These voltages deviate the carrier frequency by 1 kHz to produce two other frequencies at 2 kHz and 4 kHz. The data transmission rate is determined by the pulse width **(PW)** and the period **(PER)**. A smaller period increases the transmission

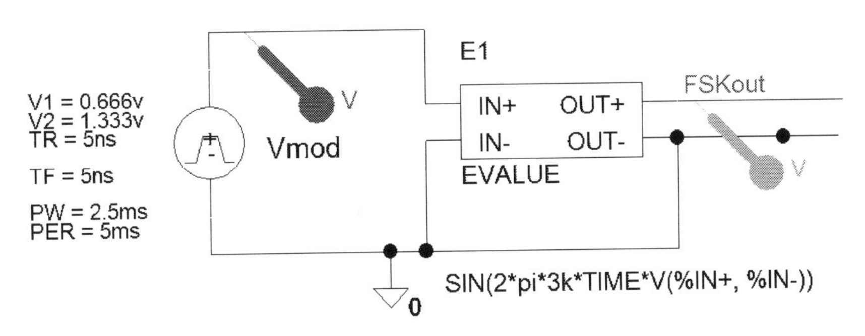

FIGURE 4.4: FSK production.

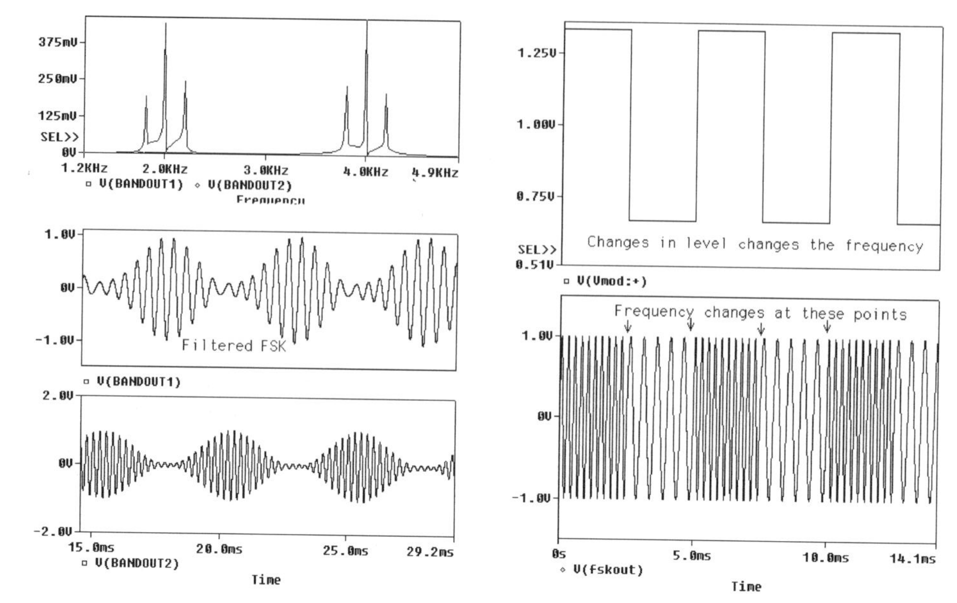

FIGURE 4.5: Modulating and VCO signals.

bit rate and frequency deviation. This increases the number of sidebands and the bandwidth. The mark and space frequencies, f_m and f_s, bit rate f_b, and modulation index h are related as

$$h = \frac{f_s - f_m}{f_b} = \frac{\Delta f}{f_b}. \tag{4.4}$$

This is similar to the modulation index, beta, in frequency modulation [ref: 10 in Appendix A]. Set the **VPULSE** generator parameters as follows: Set **Output File Options/Print values in the output file** to 40ms, **Run to time** to 50ms, and **Maximum step size** to 0.1 μs, and simulate. Fig. 4.5 shows the data and FSK signals. Select one of the variables at the bottom of the **PROBE** screen and apply the keystrokes **alt PP**, **ctrl X**, and **ctrl V** to plot the two **PROBE** signals into separate displays.

The **Plot** menu from the **PROBE** output screen contains an **Unsynchronize Plot** submenu, which enables you to display time and frequency domains simultaneously on the same screen. Increase the pulse period to 5ms and the pulse width to 2.5ms, which is a bit rate $f_b = 1/5\text{ms} = 200$ Hz. To achieve good **FFT** resolution, set Transient **Run to time** to 50ms, and simulate.

4.3.1 Frequency Shift Keying Spectrum

Fig. 4.6 shows the data and FSK signal displayed in time and frequency using the **Unsynchronize** facility from the **Plot/Unsynchronize** x-axis menu.

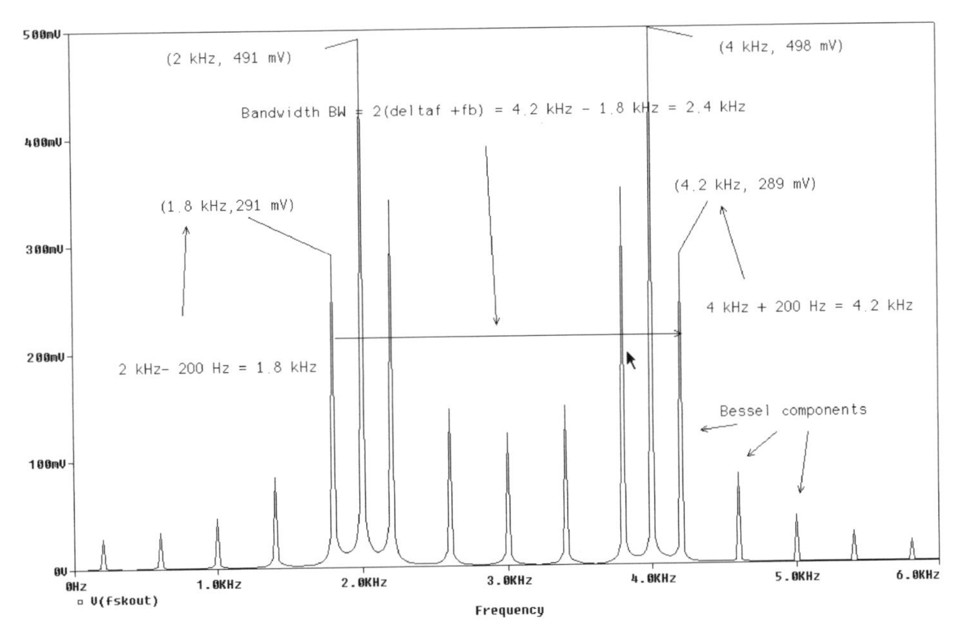

FIGURE 4.6: The data signal and corresponding FSK spectrum.

The FSK bandwidth is calculated as

$$\mathrm{BW} = 2\{\Delta f + f_b\} = 2f_b\left\{1 + \frac{\Delta f}{f_b}\right\}. \tag{4.5}$$

Substituting the mark–space frequencies into $2\Delta f = f_s - f_m$ yields the bandwidth as

$$\mathrm{BW} = 2(\Delta f + f_b) = 2(1000 + 200) = 2400\ \mathrm{Hz}. \tag{4.6}$$

The modulation index h is calculated as

$$h = \frac{f_s - f_m}{f_b} = \frac{4000 - 2000}{200} = 10. \tag{4.7}$$

4.3.2 FSK Receiver

A modern FSK demodulator uses a phase lock loop for data recovery, but here we use a non-coherent demodulating technique using two passband active filters resonant at the frequencies corresponding to the mark and space states. The specifications for the active Chebychev filter passband filter are as follows:

- The upper stopband edge frequency $f_{s2} = 3000$ Hz.
- The lower passband edge frequency $f_{p1} = 1800$ Hz.
- The upper passband edge frequency $f_{p2} = 2200$ Hz.
- The lower stopband edge frequency $f_{s1} = 1000$ Hz.
- The maximum passband attenuation $A_{max} = 0.5$ dB.
- The minimum stopband attenuation $A_{min} = 5$ dB.

The order for the filter is

$$n = \frac{\cosh^{-1}\left[\frac{10^{0.1A_{min}}-1}{10^{0.1A_{max}}-1}\right]}{\cosh^{-1}(\Omega_s)} \tag{4.8}$$

Here Ω_s is calculated as

$$\Omega_s = -\left[\frac{f_{s2}-f_{s1}}{f_{p2}-f_{p1}}\right] = \left[\frac{3000-1000}{2200-1800}\right] = \frac{2000}{400} = 5. \tag{4.9}$$

Substituting this value and A_{min}, A_{max} into (2.8), gives filter order n:

$$n = \frac{\cosh^{-1}\left[(10^{0.1\times 15}-1)/(10^{0.1\times 0.5}-1)\right]^{1/2}}{\cosh^{-1}(5)} = 0.92 \approx 1. \tag{4.10}$$

From the Chebychev loss function tables select the first-order Chebychev loss function $A(\$) = \$ + 2.863$ with 0.5 dB ripple [ref 1: Appendix A]. The bandpass frequency transform converts the low-pass loss function to a bandpass function, and denormalizes to a center frequency ω_0:

$$\$ = \frac{s^2+\omega_0^2}{Bs}. \tag{4.11}$$

The center frequency is calculated as

$$f_0 = \sqrt{f_{p1}f_{p2}} = \sqrt{1800*2200} = 2000 \text{ Hz}. \tag{4.12}$$

Expressed in radian measure as $\omega_0 = 2\pi 2000 = 12566$ r/s. The -3 dB bandwidth B is

$$B = \omega_{p2} - \omega_{p1} = 2\pi(2200-1800) = 2\pi(400) = 2512 \text{ r/s}. \tag{4.13}$$

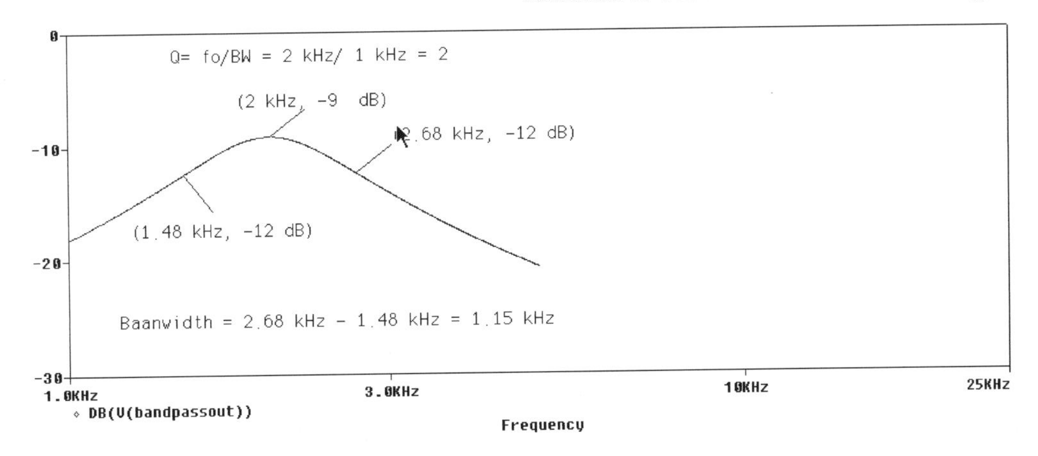

FIGURE 4.7: Bandpass response.

Substitute ω_0 and B into the frequency transform equation and then transform, yields

$$A(\$)\Big|_{\$=\frac{s^2+12566^2}{2512s}} = \left[\frac{s^2+12566^2}{2512s}\right] + 2.863. \tag{4.14}$$

The transfer function is obtained by inverting (5.14) to give

$$H(s) = \frac{1}{A(s)} = \frac{2512s}{s^2+7192.4s+12566^2}. \tag{4.15}$$

We may plot this transfer function using the **Laplace ABM** part as shown in Fig. 4.8. Numerator and denominator polynomials are entered in the Laplace part. Note the multiplying symbol "*" between the coefficient value and the complex frequency variable s. Larger numbers may be entered using the exponent system, e.g., 12566 = 1.2566e4. The transfer function is entered as follows: **NUM** = (2512*s) and the **DENOM** = (s*s + 7192.4*s+1.2566e4). From the **Analysis Setup**, select **AC Sweep** and then **Linear**, **Points/Decade** = 10000, **Start Frequency** = 1 kHz, and **End Frequency** = 3 kHz. Press **F11** to produce the bandpass frequency response in Fig. 4.7 and measure the attenuation in dB at the four specification frequencies. Change the x-axis to linear by selecting the space between the x-axis numbers. Measure the bandwidth and hence calculate the Q-factor.

4.3.3 Infinite Gain Multiple Feedback Active Filter

The infinite gain multiple feedback active filter (IGMF) in Fig. 4.8 is configured as a limited Q-factor bandpass filter for equal capacitance values $C_1 = C_2 = C$.

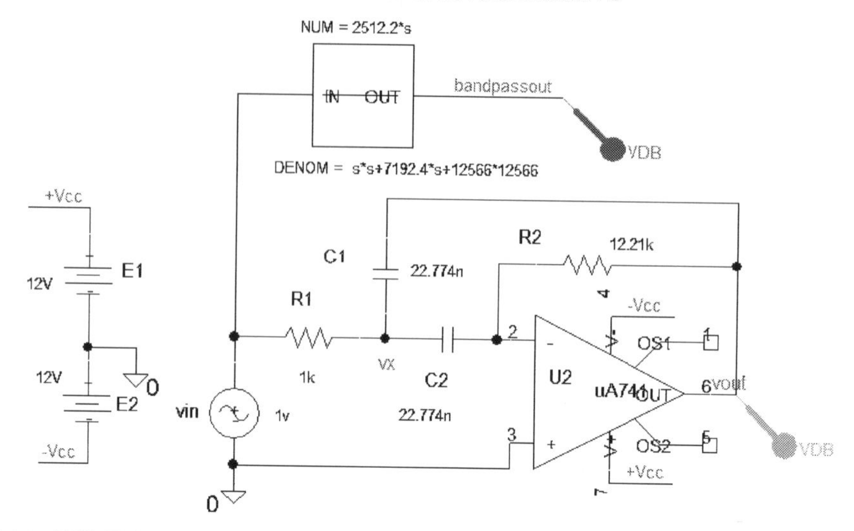

FIGURE 4.8: IGMF bandpass active filter.

The transfer function from nodal analysis at the vx junction is

$$\frac{v_{\text{out}}}{v_{\text{in}}} = -\frac{s/R_1C}{s^2 + s2/CR_2 + 1/C^2R_1R_2}. \tag{4.16}$$

Here R_2 is calculated by letting $R_1 = 1\ \text{k}\Omega$, and equating the non-s term in the denominator of (4.16) and (4.17):

$$R_2 = \frac{1}{C^2 10^3 12566^2}. \tag{4.17}$$

Eliminate the capacitor C from the s coefficient of (4.16):

$$7192.4 = \frac{2}{CR_2} \Rightarrow C = \frac{2}{R_2 7192.4}. \tag{4.18}$$

Substitute (4.19) into (4.18) to get a value for R_2:

$$R_2 = \frac{1}{\frac{2}{R_2 7192.4} 10^3 12566^2} = \frac{1}{\frac{4}{(R_2 7192.4)^2} 10^3 12566^2} = \frac{(R_2 7192.4)^2}{4 10^3 12566^2} \tag{4.19}$$

$$1 = \frac{R_2(7192.4)^2}{4 10^3 12566^2} \Rightarrow R_2 = \frac{4.10^3 12566^2}{(7192.4)^2} = 12.21\ \text{k}\Omega. \tag{4.20}$$

Substituting R_2 back into (4.18) yields the capacitance C as

$$C = \frac{2}{12210 \times 7192.4} = 22.774\ \text{nF}.$$

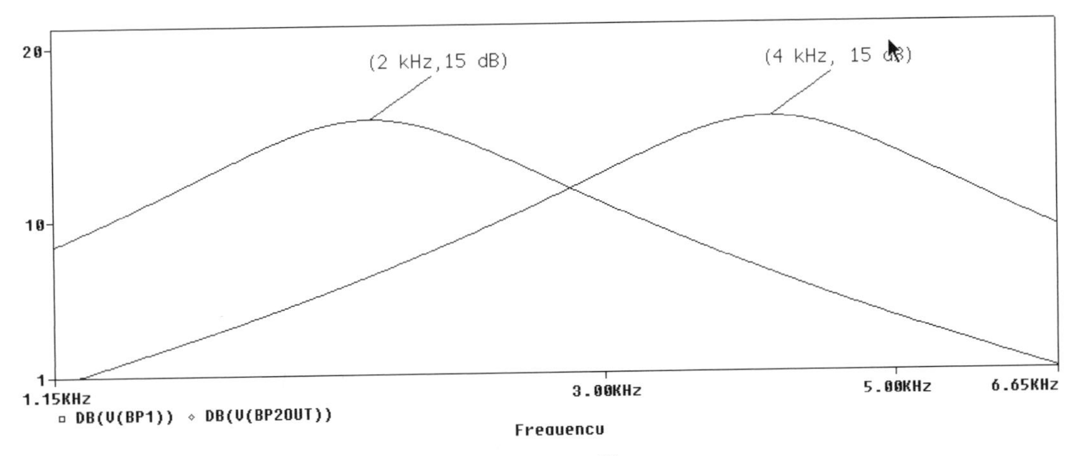

FIGURE 4.9: Passband frequency response for the two filters.

Repeat this analysis for the second bandpass filter resonant at 4 kHz. Draw the circuit using these values and from the **Analysis Setup**, select **AC Sweep/Linear, Points/Decade** = 1000, **Start Frequency** = 1 kHz, and **End Frequency** = 5 kHz. Press **F11** to simulate. The response in Fig. 4.9 is unsymmetrical around the center frequency because the Q-factor is small. The test FSK data applied to the FSK receiver, is shown in Fig. 4.10.

The FSK receiver in Fig. 4.11 is tested by generating an FSK signal using a **VPWL_F_RE_FOREVER** part that reads in a data file and applies it to an **EVALUE** part to produce an FSK test signal. This is then applied to two bandpass filters tuned to 2 kHz and 4 kHz. The output from each filter is then rectified by feeding into both inputs of a MULT part. Two **ABM** low-pass filters, with passband and stopband edge frequencies of 100 Hz and 2 kHz, respectively, to filter the doubled frequency after squaring, are then connected to a comparator to recover the original data. We have to use an ideal opamp part to satisfy the limitations of the demo version.

Set the **Output File Options/Print values in the output file** to 20 µs, **Run to time** to 100ms, and **Maximum step size** (left blank). Press **F11** key to simulate and plot the FSK data signals shown in Fig. 4.12.

4.4 PHASE SHIFT KEYING

Phase shift keying (PSK) is a form of frequency modulation called phase modulation used in deep space telemetry and cable modems. The carrier phase changes from 0–180° each time the modulating data signal changes from 0 to 1, or 1 to 0. The instantaneous value for a PSK signal is

$$v_c(t) = E_c \cos(2\pi f_c t + \theta(t))\ \mathrm{V}. \tag{4.21}$$

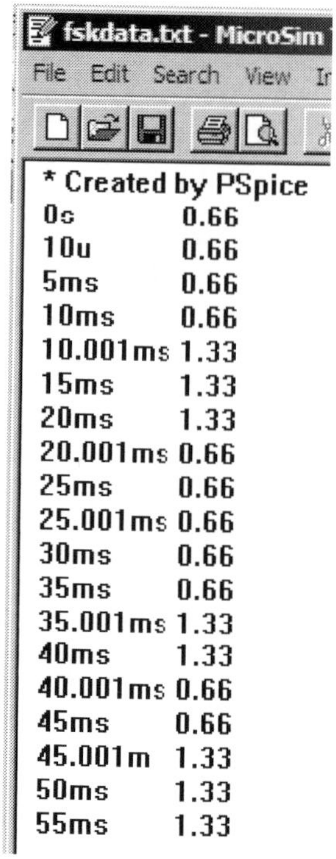

```
fskdata.txt - MicroSim
File Edit Search View Ir
* Created by PSpice
0s          0.66
10u         0.66
5ms         0.66
10ms        0.66
10.001ms 1.33
15ms        1.33
20ms        1.33
20.001ms 0.66
25ms        0.66
25.001ms 0.66
30ms        0.66
35ms        0.66
35.001ms 1.33
40ms        1.33
40.001ms 0.66
45ms        0.66
45.001m   1.33
50ms        1.33
55ms        1.33
```

FIGURE 4.10: FSK data applied to the **VPWL_F_RE_FOREVER** part.

The phase $\theta(t)$ is zero or π radians and changes whenever the data changes from $+1$ to -1. The carrier phase thus depends on the digital input signal $d(t) = \pm 1$. Rewrite (4.21) as

$$v_c(t) = d(t) E_c \cos(2\pi f_c t) \text{ V}. \tag{4.22}$$

Draw the PSK schematic shown in Fig. 4.13.

Center-tapped transformers use the **K_Linear** part to link the inductors together by specifying the coupling and the coil names. Select this part and enter the inductor names and the coupling coefficient. Be careful about placing inductors on the schematic area. The inductor might not have the dot on the symbol. If it has no dot on the symbol, then an imaginary DOT must be imagined at the top of the symbol when placed initially on the schematic area. Rotating the coil 180° produces 180° phase shift in any applied signal with the DOT at the bottom. Set the transformer parameters as in Fig. 4.14, by selecting the boxed **K** part. Data is simulated using a square wave **VPULSE** generator part with parameters V1 = 1 V, V2 = −1 V, $t_r = t_f = 1\mu$s, **PW** = 2ms, **PER** = 4ms. The fundamental frequency of the 4ms data is 250 Hz. Set **Run to**

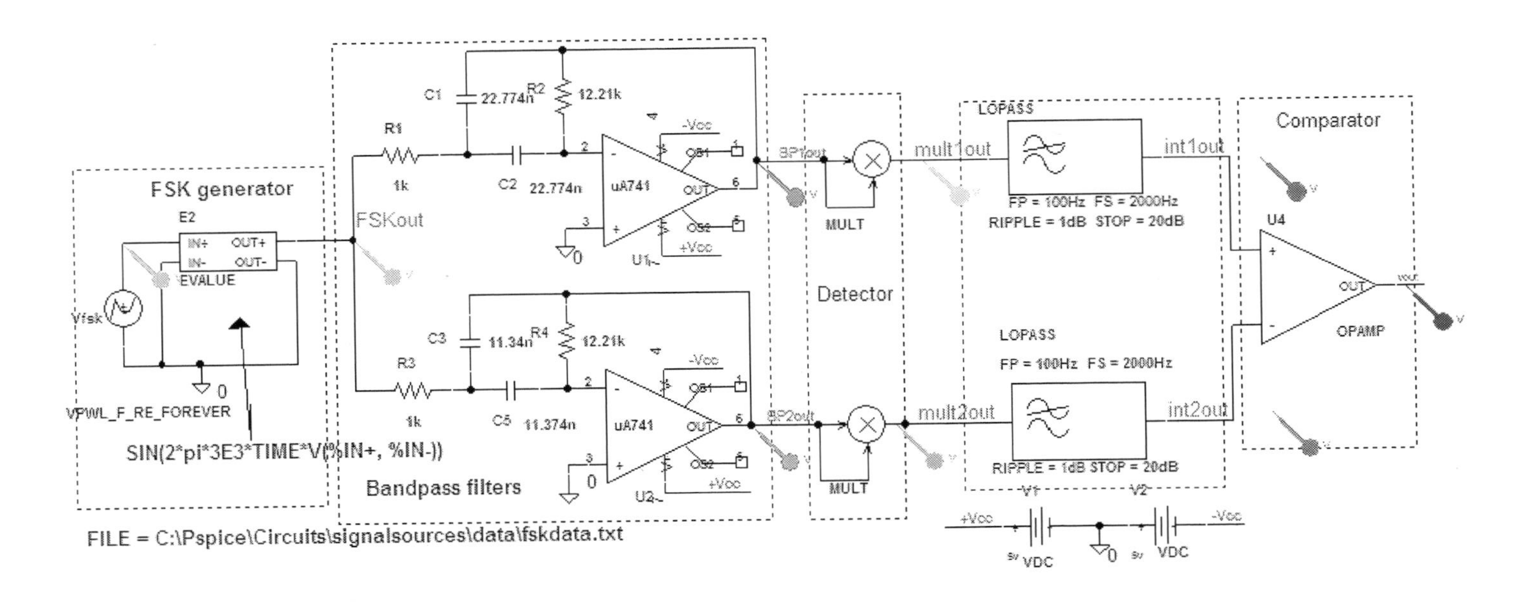

FIGURE 4.11: FSK receiver.

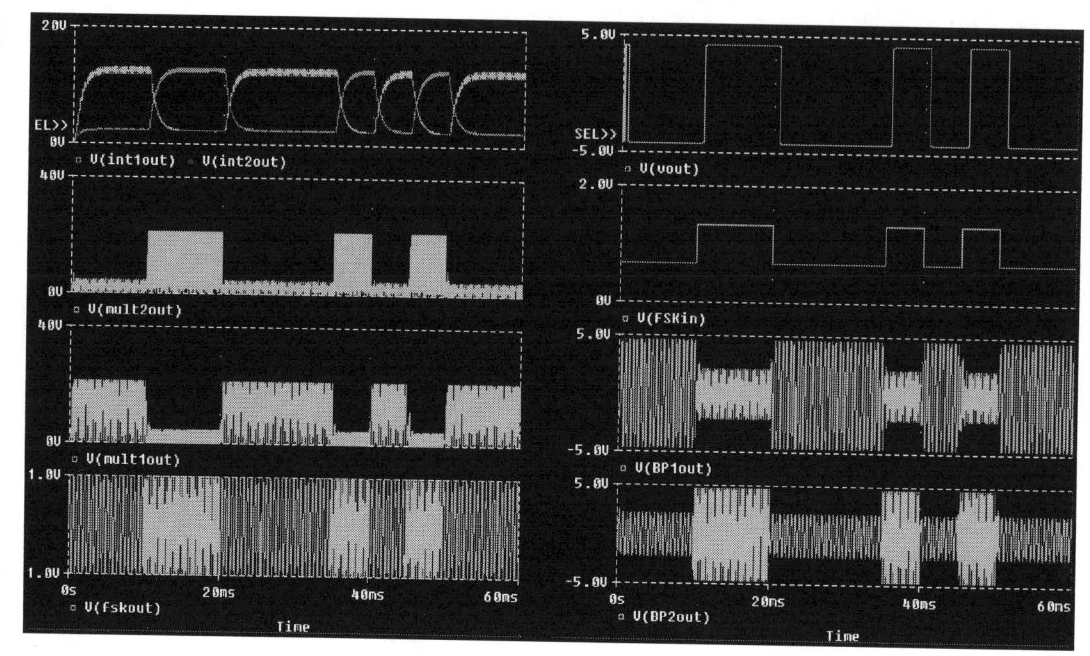

FIGURE 4.12: Time waveforms for FSK.

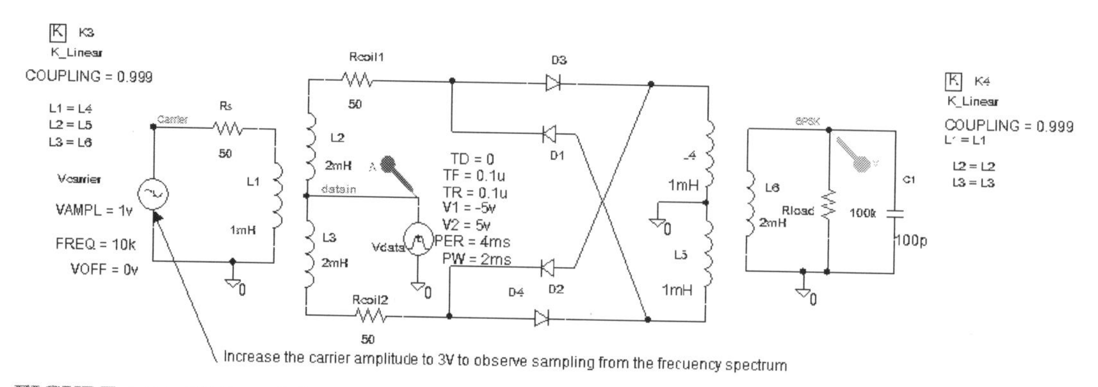

FIGURE 4.13: PSK modulator.

time to 15ms and **Maximum step size** to 1 μs. Press **F11** to simulate. Use the magnifying tool to reduce the time axis to 4ms and obtain the carrier phase shift in Fig. 4.14 using the cursors to measure the time difference Δt, and converting it to phase using (4.23):

$$\theta = \frac{360 \times \Delta t}{T}. \tag{4.23}$$

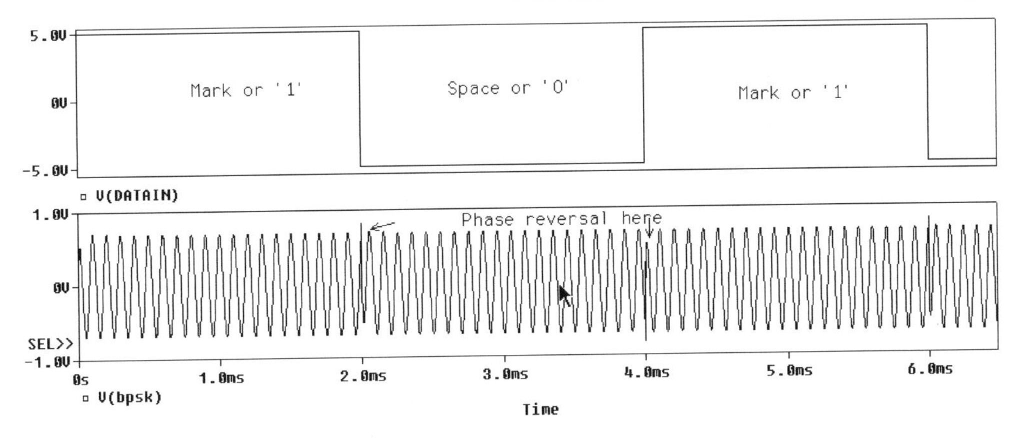

FIGURE 4.14: PSK signal and the data signal.

Here T is the period of the signal (0.1ms for the 10 kHz carrier). Thus, the phase shift is

$$\theta = \frac{360^\circ \times 50\ \mu\text{s}}{100\ \mu\text{s}} = 180^\circ. \tag{4.24}$$

Select the **FFT** icon in **PROBE** to display signals in the frequency domain as shown in Fig. 4.15. Frequency resolution is determined by the **Run to time** parameter in the **Transient Setup**. Increasing the **Run to time** achieves greater **FFT** resolution but at a cost of increased simulation time.

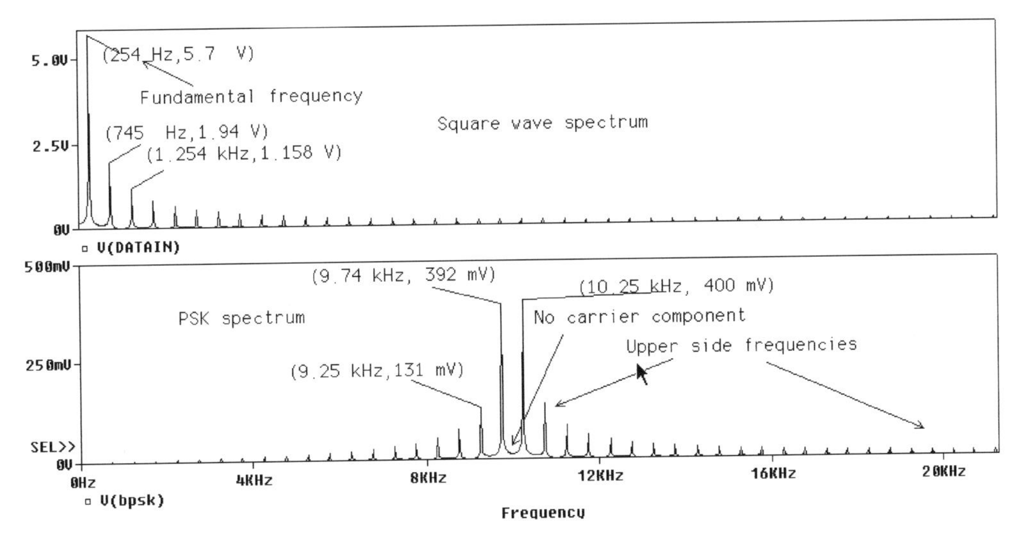

FIGURE 4.15: PSK spectrum.

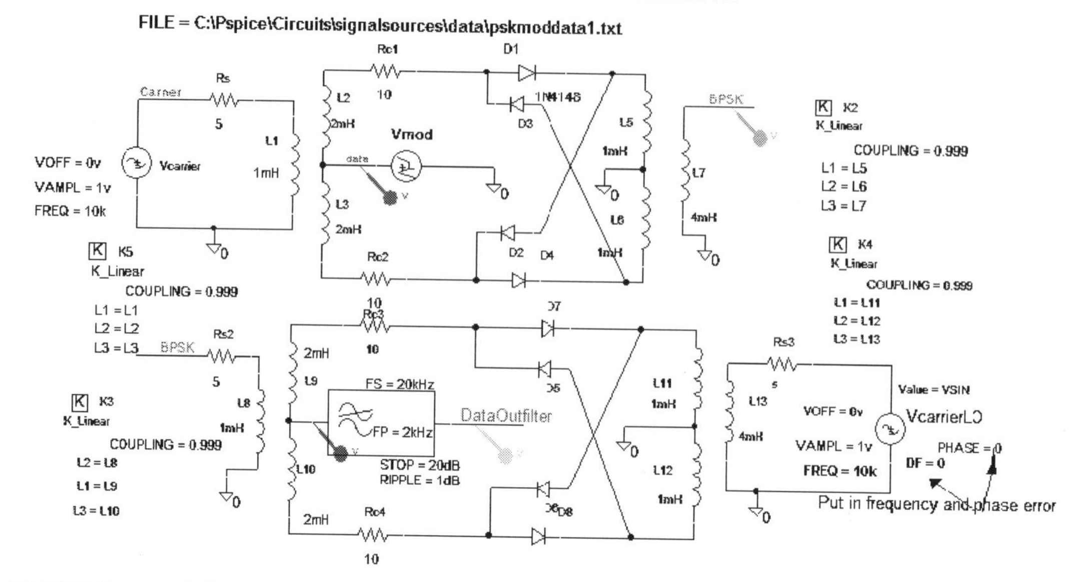

FIGURE 4.16: PSK demodulator.

Increase the carrier amplitude to 3 V and change the data frequency to 1 kHz to modulate the 10 kHz carrier signal. The side frequencies are repeated at multiples of the carrier frequency (this is a sampled signal spectrum). The spectral components are located at 9 kHz and 11 kHz and repeated at 29 kHz and 31 kHz (a third harmonic of the carrier). The bandwidth, BW, is obtained by subtracting the lowest side frequency from the upper side frequency:

$$\mathrm{BW} = (f_c + f_m) - (f_c - f_m) = 2f_m = 2\ \mathrm{kHz}. \tag{4.25}$$

The carrier is not in the spectrum and it is necessary in the receiver to generate a carrier with an identical frequency and approximately the same phase as the transmitted carrier—a process called coherent carrier generation.

4.4.1 PSK Receiver

The PSK demodulator shown in Fig. 4.16 contains balanced modulators (BM) to recover the data from the PSK signal by multiplying the incoming bandpass PSK signal by an identical carrier:

$$v_c(t) = \mathrm{BPSK}\cos(2\pi f_c t) = d(t)E_c \cos(2\pi f_c t)\cos(2\pi f_c t). \tag{4.26}$$

Expand (4.26) using cos A cos B = 0.5[cos(A+B) + cos(A−B)] as $v_c(t) = 0.5d(t)E_c \cos(4\pi f_c t)$.

The PSK output wire segment is called BPSK and connects to the BM input because it is also called BPSK. The local oscillator **VcarrierLO** is connected as shown and can have errors

```
pskinput.txt - N
File  Edit  Format
0us,        0
2ms,        0
2.001ms,    5
4ms,        5
4.001ms,    0
6ms,       -5
6.001ms,    5
```

FIGURE 4.17: The test data signal "pskmoddata1.txt."

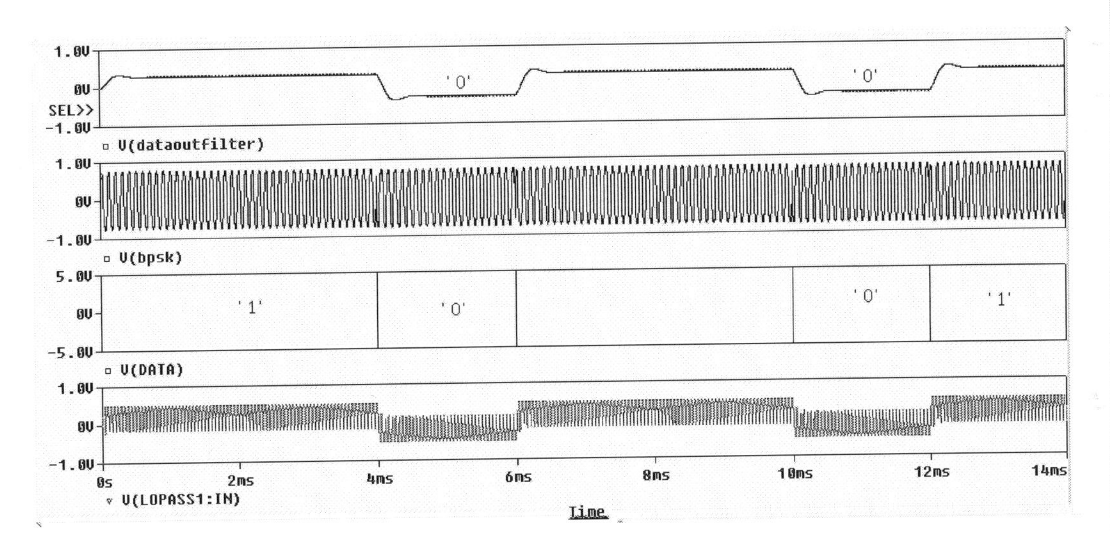

FIGURE 4.18: Recovered data.

in frequency and phase placed to simulate the non-coherent conditions. The modulating data generator is replaced by a random signal applied using the **VPWL_F_RE_N_TIMES** generator part that reads in a text file, as shown in Fig. 4.17. The file is called "pskinput.txt," and is located at C:\Pspice\Circuits\signalsources\data\pskmoddata1.txt.

The recovered data output signal from the center-tapped transformer uses a low-pass filter **ABM** part called **LOPASS** to remove unwanted high-frequency components. Set **Output File Options/Print values in the output file** to 15ms, **Run to time** to 20ms, and **Maximum step size** to 1 μs. Simulate and observe the **PROBE** output signals in Fig. 4.18.

TABLE 4.1: XNOR Truth Table

A	B	OUT
0	0	1
0	1	0
1	0	0
1	1	1

Investigate the problems associated with a noncoherent local oscillator (meaning not the same frequency or phase as the carrier transmitted), by varying the frequency and phase of the local carrier oscillator VcarrierLO.

4.5 DIFFERENTIAL PHASE SHIFT KEYING (DPSK)

The difficulty of generating a local carrier in a receiver, with the same frequency and phase as the transmitted carrier (called a coherent carrier), is overcome using differential phase shift keying modulation [ref: 4 Appendix A]. The truth table for the data signals applied to the XNOR gate is shown in Table 4.1.

The B input is initially assumed 0 and is compared to the first data bit, which is "1". This produces a 0 at the gate output, because both inputs are different. This 0 is then delayed by one bit and fed back to the B gate input, where it is compared to the second inputted bit (a 0), to produce a 1 at the output. Table 4.2 shows the logic values at different points in the circuit. The one bit delay is achieved using a clocked D-type flip-flop, where the data is shifted through on the low-to-high clock pulse transitions.

The NRZ signal at the gate output is converted to an NRZ-B signal by a comparator whose output is ±4 (the polarity depends on the value of the digital input). The comparator

TABLE 4.2: PSK Data Signals

DATA IN A	1	0	0	1	0	1	1	1
B (delay) [0]	0	0	1	0	0	1	1	1
NRZ	0	1	0	0	1	1	1	1
NRZ-B	−1	1	−1	−1	1	1	1	1
Carrier phase	π	0	π	π	0	0	0	0

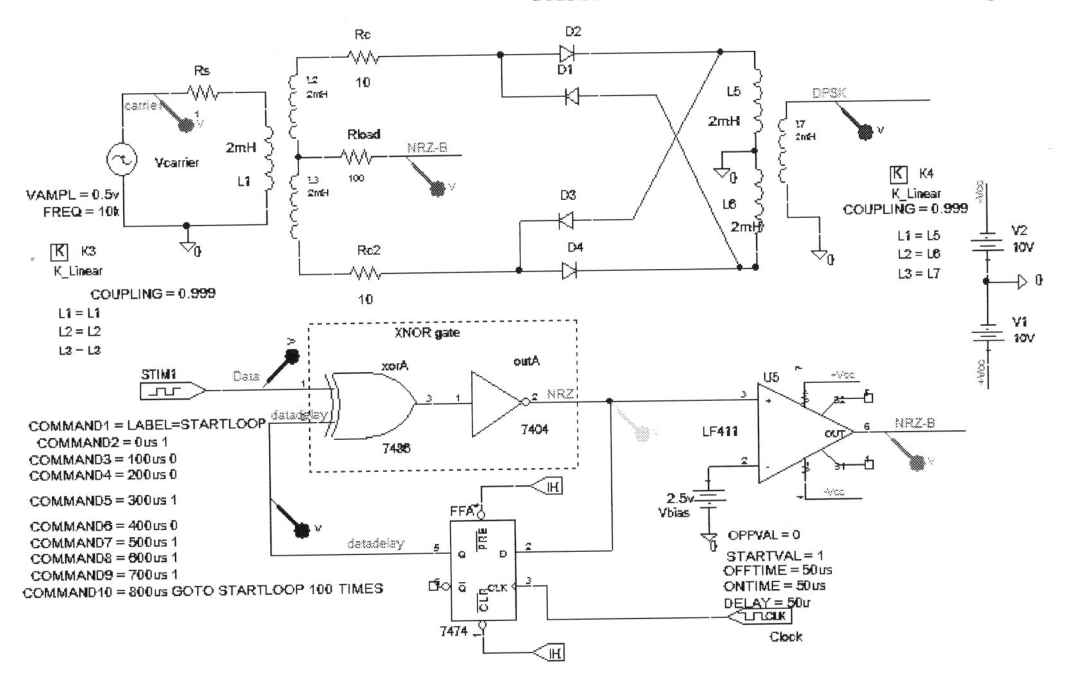

FIGURE 4.19: DPSK modulator.

output saturates to the positive rail voltage for a +1 V input signal, but saturates to the negative rail voltage when the positive input is zero because of the +V/2 at the inverting input. The NRZB signal switches the pairs of diodes (*D*1, *D*3) and (*D*2, *D*4) on or off, depending on the signal polarity. The signal path is reversed if −1 is present at the center tap, but no path inversion for +1. The instantaneous amplitude of a carrier modulated by the data is

$$v_c(t) = E_c \cos(2\pi f_c t + \theta(t)) \text{ V}. \tag{4.27}$$

The phase change $\theta(t)$ is 0, or π, depending on the sign of the modulating digital input signal ±1. Equation (4.27) is rewritten as

$$v_c(t) = d(t) E_c \cos(2\pi f_c t) \text{ V}, \tag{4.28}$$

where $d(t) = \pm 1$. Fig. 4.19 shows a balanced modulator with the center-tapped transformer formed using a **K_Linear** part. Select the **K** part (enclosed by a box) and enter the inductor part names as shown.

The XNOR is a two-input XOR gate and an inverter combined because the evaluation version does not have an XNOR part. The NRZ-B output is applied to the center tap via a resistance to prevent loading the LF411 output when the diodes conduct. Set the data signal **STIM1** generator part parameters as shown. The **DigClock** is set with equal on and off

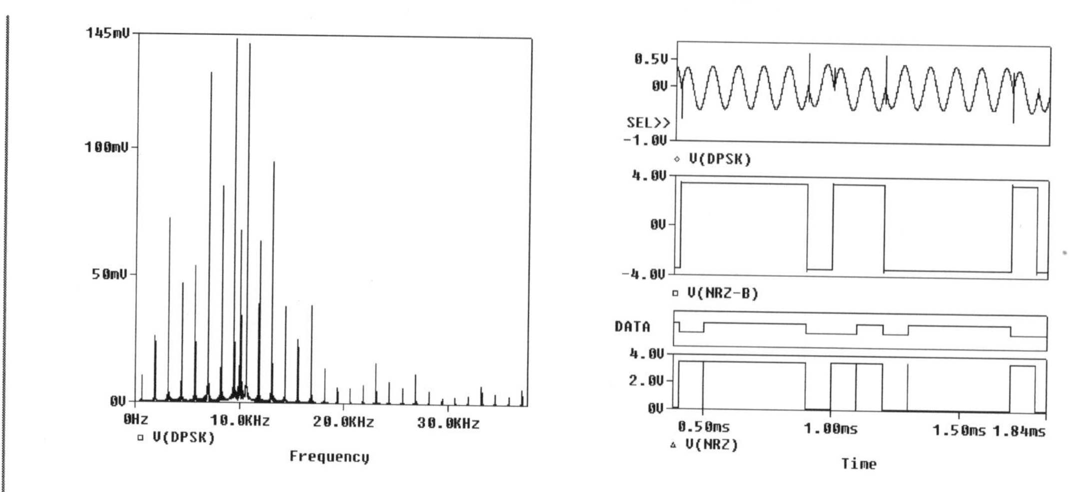

FIGURE 4.20: NRZ-B, NRZ, and DPSK output carrier signals.

time = 50 μs (the inverse of the 100 kHz clock frequency). Set the **Output File Options/Print values in the output file** to 99ms, **Run to time** to 100ms, and **Maximum step size** to 0.1 μs. Simulate with the **F11** key to show the data, delayed data, and clock signals as in Fig. 4.20.

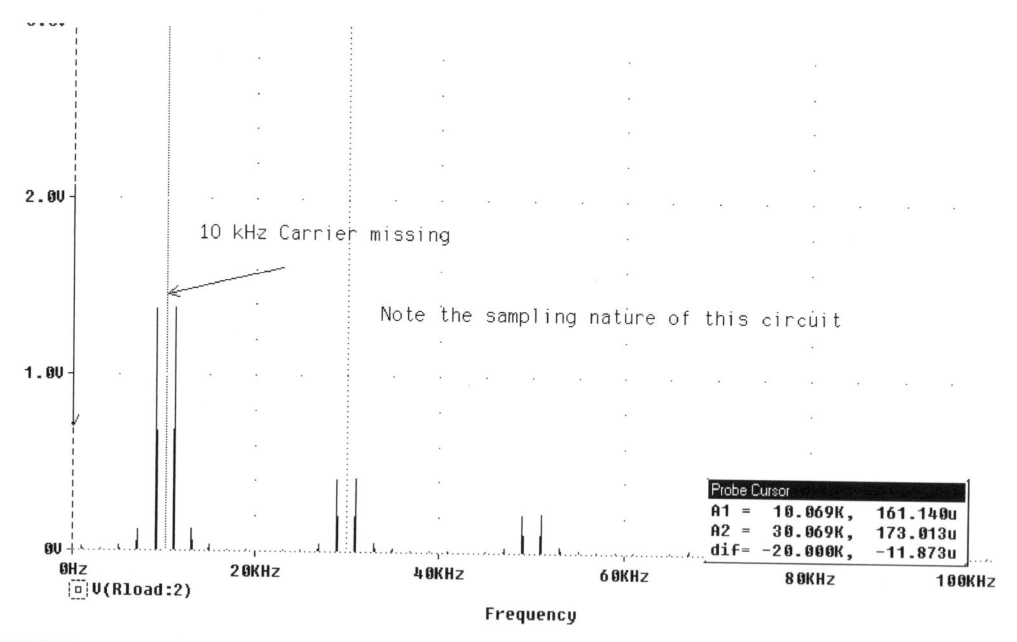

FIGURE 4.21: DPSK spectrum showing the sampled nature.

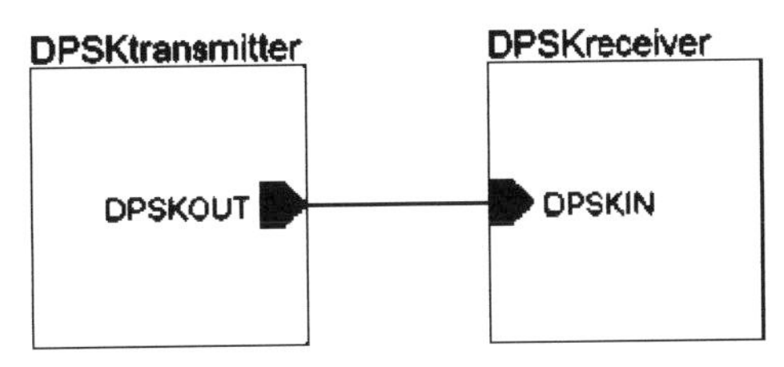

FIGURE 4.22: Complete DPSK system.

The **FFT** icon displays the signal in the frequency domain. For greater frequency resolution, but a longer simulation time, increase **Run to time**. The spectral components for the 10 kHz carrier signal modulated by a 1 kHz modulating signal are repeated at multiples of the carrier because it is a sampling process. The DPSK bandwidth is calculated by subtracting the lowest side frequency from the highest side frequency:

$$\mathrm{BW} = (f_c + f_m) - (f_c - f_m) = 2f_m = 2\ \mathrm{kHz}. \tag{4.29}$$

4.6 DIFFERENTIAL PHASE SHIFT RECEIVER

In Fig. 4.22, we use the hierarchical system to create a DPSK transmission and receiver system into manageable blocks. The first block "DpskTransmitter" is the transmitter examined in the previous section.

The received signal and a delayed version of it (use a **T** part to delay the signal) are fed to a **MULT** part, which acts as a phase detector (PD). The PD output is fed into an **ABM** low-pass filter to remove the product term, and the signal in the comparator converts the NRZ-B signal to a TTL NRZ signal format. To see why a low-pass filter (LPF) is necessary, assume that the input is $\sin(x)$, so that the delayed version is also $\sin(x)$:

$$\sin(x)\sin(x) = \sin^2(x) = 0.5(1 - \cos 2x). \tag{4.30}$$

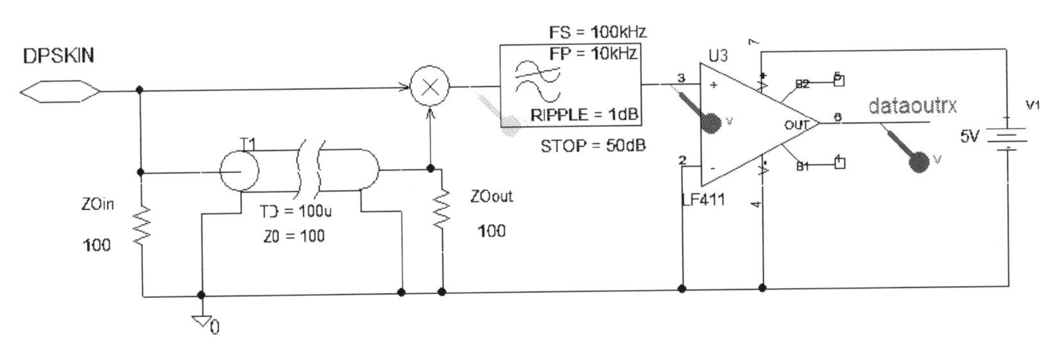

FIGURE 4.23: DPSK receiver.

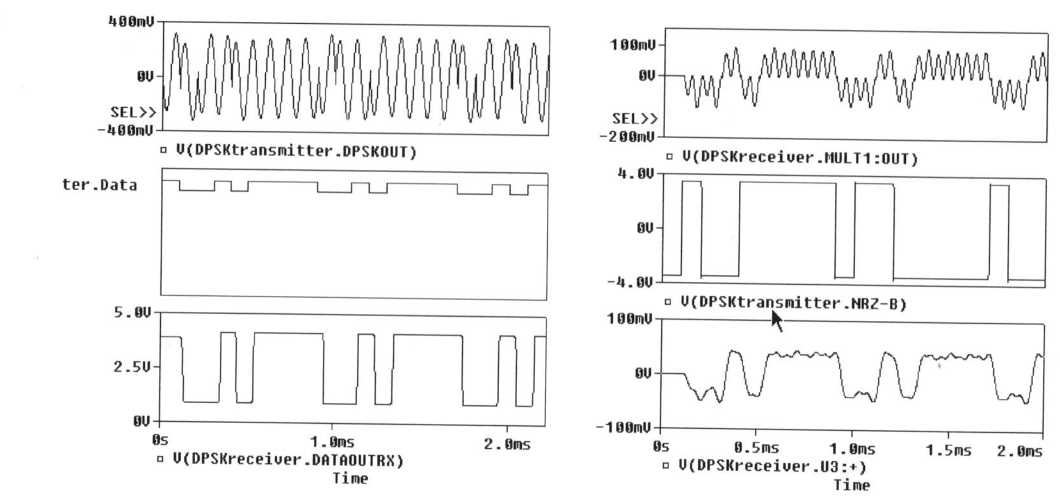

FIGURE 4.24: DPSK receiver signals.

The LPF filters the $\cos(2x)$ term with a 0.5 V DC term remaining. However, if a phase reversal occurs, the inputs to the phase detector are $A\sin(x)$ and $-A\sin(x)$, thus $\sin(x)(-\sin(x)) = -1/2(1 - \cos(2x)) = -1/2$ (see QPSK discussion further on in this chapter). The receiver schematic in Fig. 4.23 uses a **T** part transmission line to achieve the 100 μs delay. However, the transmission line must be correctly terminated at the input and output ports, with resistances equal to the characteristic impedance. The **ABM** filter parameters are as follows: **Passband Frequency** = 10 kHz, **Stopband Frequency** = 100 kHz, **Ripple** = 1 dB, and **Stopband Edge Attention** = 50 dB.

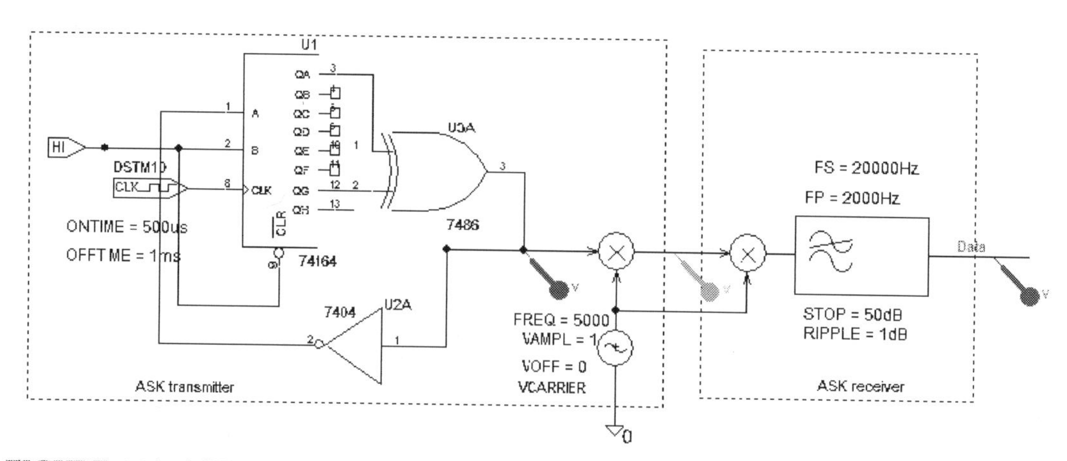

FIGURE 4.25: ASK modem using a stimulus part clock.

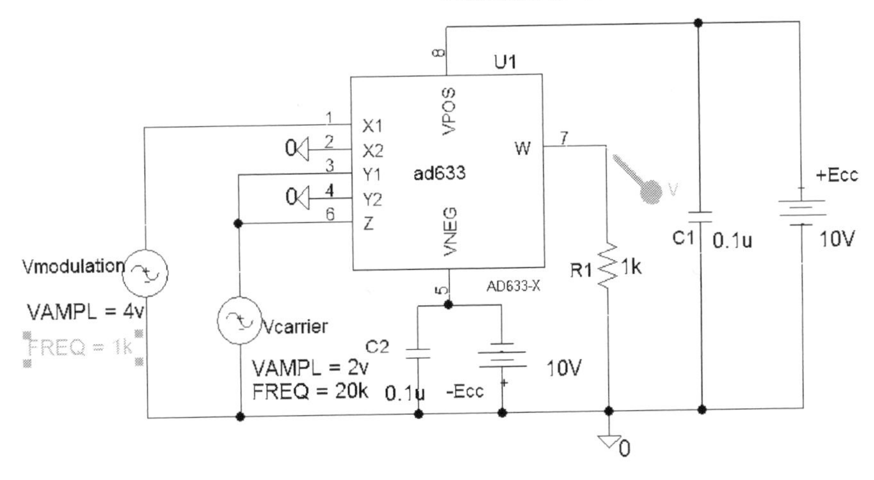

FIGURE 4.26: ASK using the AD633.

Set **Output File Options/Print values in the output file** to 1ms, **Run to time** to 8ms, and **Maximum step size** to 1 μs, and simulate with the **F11** key. The recovered data signal is shown in Fig. 4.24. Separate the displays and use the magnifying tool to examine the signal in detail. Consult the index for an explanation of the log command for automating the plot separation process.

4.7 EXERCISES

1. Investigate the ASK modem in Fig. 4.25. This schematic generates the input data that uses a PRBS generator and **ABM** parts.
2. Investigate amplitude shift keying (ASK) using the IC multiplier AD633 shown in Fig. 4.26. Make the transient **Run to time** equal to 4ms.

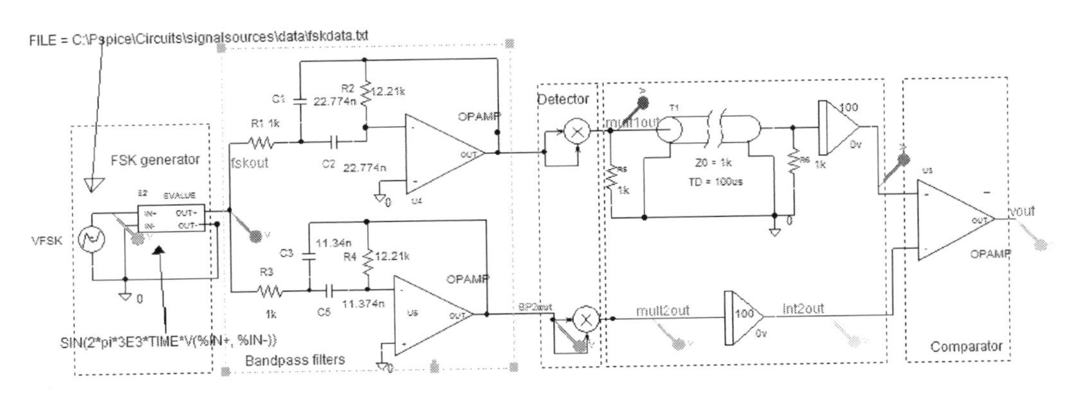

FIGURE 4.27: FSK receiver 2.

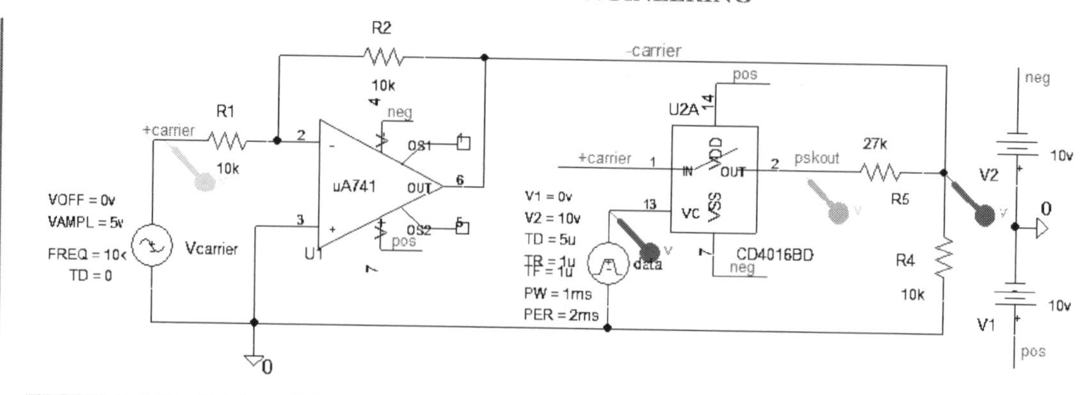

FIGURE 4.28: PSK modulator.

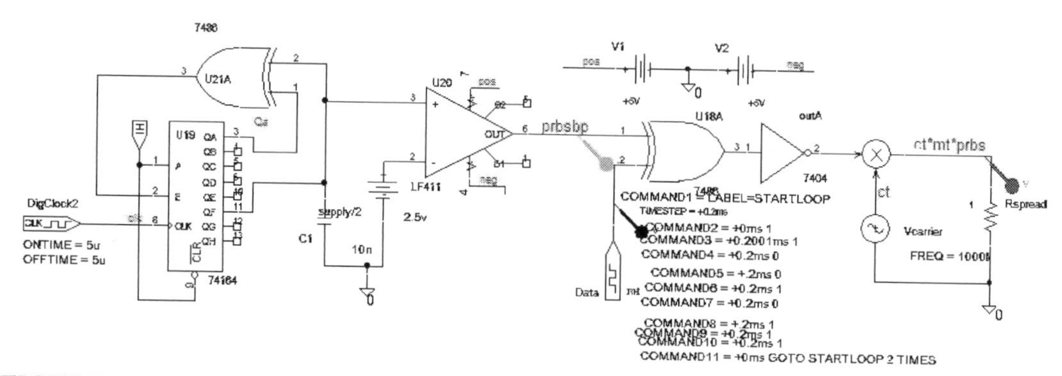

FIGURE 4.29: Digital multiplier using an XNOR gate.

3. Investigate the FSK receiver in Fig. 4.27. Note the use of the **MULT** to produce a squaring function and a correctly terminated transmission line for a delay.

   Should the integrators have some sort of switching mechanism as the "integrate and dump" circuit investigated previously (see index)?

4. Use two **ABM** phase-locked loops to recover the data from an FSK signal.
5. Fig. 4.28 shows a PSK circuit that uses a CD4016 IC switch. Investigate this circuit and measure the passband bandwidth (BW).
6. Investigate PSK modulation using the AD633 multiplier.
7. Investigate the two digital sequences applied to the XOR gate as shown in Fig. 4.29. A simple pseudorandom bit sequence (PRBS) generator is used to provide one of the data streams. A modulator is also shown, but this is investigated in a later chapter on spread spectrum systems.

CHAPTER 5

Multilevel Signaling and Bandwidth Efficiency

5.1 CHANNEL CAPACITY

The Shannon–Hartley theorem develops an expression for the maximum transmission rate called the channel capacity, C, through a communication channel and is defined as

$$C = \frac{\text{information}}{\text{total message time}} = \frac{H}{T_m}. \tag{5.1}$$

The information transmitted, for a total message time, T_m, is

$$H = C T_m. \tag{5.2}$$

In Fig. 5.1, there are M levels (M-ary) displayed, where a level may be a voltage, current, or phase change in the signal. We will examine shortly a passband modulation technique called quadrature phase shift keying (QPSK); so in this four-level system, each level is one of four phase shifts in the carrier. The information time slot is T_b for the total message time T_m. For example, in QPSK assume each phase shift is likely to occur, then in the first time slot there are four phase and in the second time slot there are another four phase giving a total of $4 \times 4 = 4^2 = 16$ [ref: 4 Appendix A]. In general, we can write

$$M^{T_m/T_b}. \tag{5.3}$$

Taking the logarithm of (5.3),

$$\log_2(M^{T_m/T_b}) = T_m/T_b \log_2 M. \tag{5.4}$$

This quantity is proportional to the amount of information, so dividing it by the message signal duration gives the channel capacity as

$$C = (T_m/T_b \log_2 M)/T_m = 1/T_b \log_2 M. \tag{5.5}$$

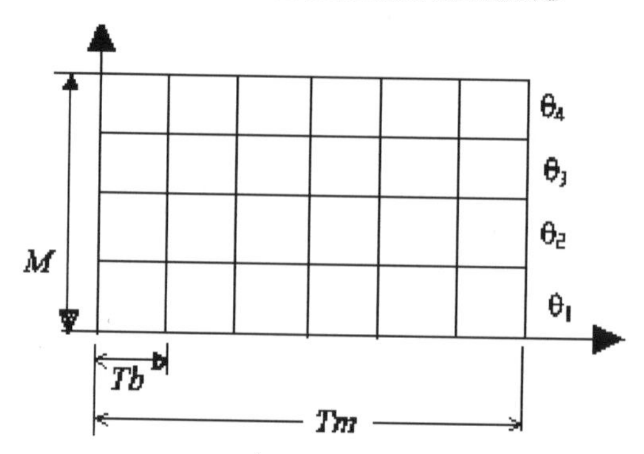

FIGURE 5.1: M–T_m space.

The absolute minimum bandwidth BW is $f_{c\,\min} = 1/2T_b$, so substituting for $1/T_B = 2f_{c\,\min}$ yields

$$C = 2f_{c\,\min} \log_2 M. \tag{5.6}$$

A binary message has two levels, i.e., $M = 2$, so C is

$$C = 2f_{c\,\min} \log_2 2 = 2f_{c\,\min} = 2B. \tag{5.7}$$

5.2 MULTILEVEL ENCODING: BANDWIDTH EFFICIENCY

Higher bit rates may be transmitted over a finite bandwidth channel using multilevel (M-ary) signaling. Noise, however, prevents levels from being too close because it would be impossible to distinguish between adjacent levels. A large number of M levels require minimum spacing, maximum pulse amplitudes, and large signal power. Claude Shannon, in his famous 1948 paper, showed how the maximum possible transmission rate, or channel capacity, for a channel, with bandwidth B Hz and signal to noise ratio S/N, is

$$C = B \log_2 \left\{\frac{S+N}{N}\right\} = B \log_2 \left\{1 + \frac{S}{N}\right\} \text{ bps.} \tag{5.8}$$

This theoretical limit is not easily obtained because increasing the bandwidth also increases the amount of noise, so there is an upper Shannon limit of 1.44 times the signal to noise ratio. A channel with bandwidth B has a maximum signaling rate of $2B$ pulses per second, so for $B = 1$ Hz we can transmit two pulses per second. This is valid only if the channel has an ideal flat frequency response, no distortion, no noise present, and the data has a specific pulse shape (a very tall order!!). The bandwidth determines the highest pulse rate at which intersymbol interference (ISI) is avoided. For M-ary signaling, each level contains $\log_2 M$ bits of information, so the maximum bit rate is $2B\log_2 M$ bps. We need to introduce a parameter to measure how well a particular modem uses the available bandwidth. This is called the bandwidth efficiency (also called information density), and it compares digital systems with the efficiency normalized to a bandwidth of 1 Hz. The bandwidth efficiency, in bits/second/Hz, is the ratio of maximum bit rate C to the bandwidth B and has a maximum value

$$\frac{C}{B} = \frac{\text{Transmission rate (bps)}}{\text{Minimum bandwidth (Hz)}} = 2\log_2 M \text{ bits/Hz.} \tag{5.9}$$

In theory, we achieve high bit rates from a channel with a finite bandwidth using multilevel signaling, with each pulse having different levels (up to 1024 levels with some modems). The first technique splits data bits into pair-bits called dibits, i.e., 00, 01, 10, and 11, with each dibit assigned one of four different amplitudes and sent as a symbol (tribits are formed from three-bit groups). The symbol may be a voltage, current, or a phase change. This produces a transmission rate that is different to sending data in binary format where each pulse carries one bit of information and is called the bit rate. The transmission rate for dibits and tribits is expressed in baud, which is the signaling or symbol rate. The baud rate is therefore the bit rate divided by the number of bits per symbol. The number of levels M, for n bits, is

$$M = 2^n \Rightarrow n = \log_2 M. \tag{5.10}$$

QPSK is an example of M-ary signaling used in modem (**mo**dulator/**dem**odulator) technology using four phase states as in CDMA systems and has very good power efficiency and noise immunity. Noise prevents the levels from being too close together as it would be impossible to distinguish between adjacent levels. A minimum level spacing and large maximum pulse amplitudes (requires large signal power) are required when the number of levels is large. High data rate transmission is not possible for a noisy finite bandwidth channel and fixed signal power.

5.3 BIT ERROR RATE

A probability of bit error occurring—the bit error rate (BER)—is a measure of the number of errors detected in a sequence length from the incoming transmitted data in a receiver. BER rates 10^{-3} to 10^{-9} (1 error per 1000 sent to 1 error per 10^9 sent) are a function of the signal to noise ratio, SNR. However, for digital signals it is usual to redefine the SNR as an $E_b N_0$ ratio. The average signal power, S W, is measured at the receiver input, where each bit takes $1/C$ s to transmit, with each bit having an average energy per bit $E_b = S/C$ W s, or J (joules) (power, the rate of expenditure of energy, is energy/time). The signal to noise power ratio in a channel with total noise power $N_0 B$ W is

$$\text{SNR} = \frac{E_b}{N_0}\frac{C}{B}. \tag{5.11}$$

The signal to noise power ratio in terms of bandwidth efficiency C/B and $E_b N_0$ ratio is

$$\frac{S}{N} = \frac{E_b}{N_0} * [\text{Bandwidth efficiency}]. \tag{5.12}$$

Thus, the system SNR, for a given bandwidth efficiency, is proportional to E_b/N_0 and hence the BER. A multilevel digital channel capacity is

$$C = B \log_2 \left\{ 1 + \frac{E_b R}{N_0 B} \right\} \text{ bps}. \tag{5.13}$$

A symbol error rate (or SER) also defines the error rate for symbols as in M-ary signaling.

5.4 QUADRATURE PHASE SHIFT KEYING

Quadrature phase shift keying (QPSK) is an M-ary level modulation scheme used in satellite communication systems where a carrier is switched into one of four phase states. The QPSK modem in Fig. 5.2 shows two multipliers (balanced modulators) to switch the carrier path for positive and negative data signals. We investigated PSK in Section 4.4 where the modulating data caused pairs of diodes to switch the carrier path to give the required phase shift. Here two orthogonal signals are generated: I-channel (in-phase) signal and a Q-channel (quadrature) signal. We split the data into dibit pairs using a bit-splitter, where the relationship between the phase and the dibits is shown in Table 5.1. Consider the dibit pair $a_1 = 0$, $b_1 = 1$: the b_1 signal reverses the signal path of the carrier 90° vector, i.e., −90°, but the a_1 signal causes no signal path change. The total output phase is the addition of the 0° and −90° vectors and results in a total phase change of −315°. The dibits are gray-coded with only one bit changing in every dibit per successive phase change because it produces a better BER performance. The

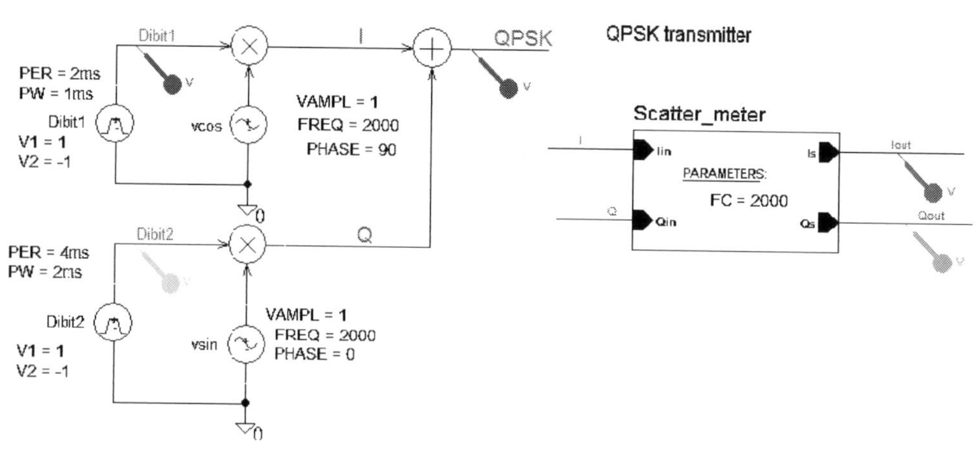

FIGURE 5.2: QPSK signal generation using **ABM** parts.

TABLE 5.1: Dibit Pair–Phase Relationships

A	B	A	B
00	10	11	01
45^0	135^0	225^0	315^0

instantaneous value of a 10 kHz QPSK carrier is

$$V_{\text{QPSK}} = \cos\left[2\pi 10^3 t + \frac{\pi}{2}(1 + d_1(t))\right] + \sin\left[2\pi 10^3 t + \frac{\pi}{2}(1 + d_2(t))\right]. \qquad (5.14)$$

The magnitude of the path switching functions $d_1(t)$ and $d_2(t)$ is $|d(t)| = \pm 1$ and Table 5.1 shows the phase for each dibit pair.

5.5 QPSK MODULATION USING ABM PARTS

The final QPSK output is the sum of the orthogonal vectors. The "00" dibit has a phase of 45°, so the magnitude for each **VECTOR** is

$$|V_{\text{QPSK}}| = \sqrt{\cos(\pi/4) + \sin(\pi/4)} = \sqrt{0.707 + 0.707} = 1. \qquad (5.15)$$

Each carrier has a magnitude of 1 V and frequency of 2000 Hz. The scatter meter is examine in chapter 6 but here we introduce it to examine the constellation formed from the quadrature signals. The scatter meter displays the dibit pairs in a constellation diagram and to use the meter, name the wire segments at each multiplier output with the same names as shown on the scatter meter input ports.

Set **Output File Options/Print values in the output file** = 1ms, **Run to time** = 10ms, and **Maximum step size** = 1 µs. Press the **F11** key to produce the QPSK signal shown in Fig. 5.3.

In Fig. 5.4, magnifying a section of the carrier plot enables us to measure the carrier phase change by placing one cursor at the maximum carrier value and moving the second cursor to the next phase change value and measure the time difference between the two cursors as 187 µs. Multiplying this time by 360 and the carrier frequency yields the angle corresponding to the dibit pair "10":

$$\theta = \frac{360 \times 187u}{T_c} = 360 \times 187 \times 10^{-6} fc = 360 \times 187 \times 10^{-6} \times 2000 = 135^\circ. \qquad (5.16)$$

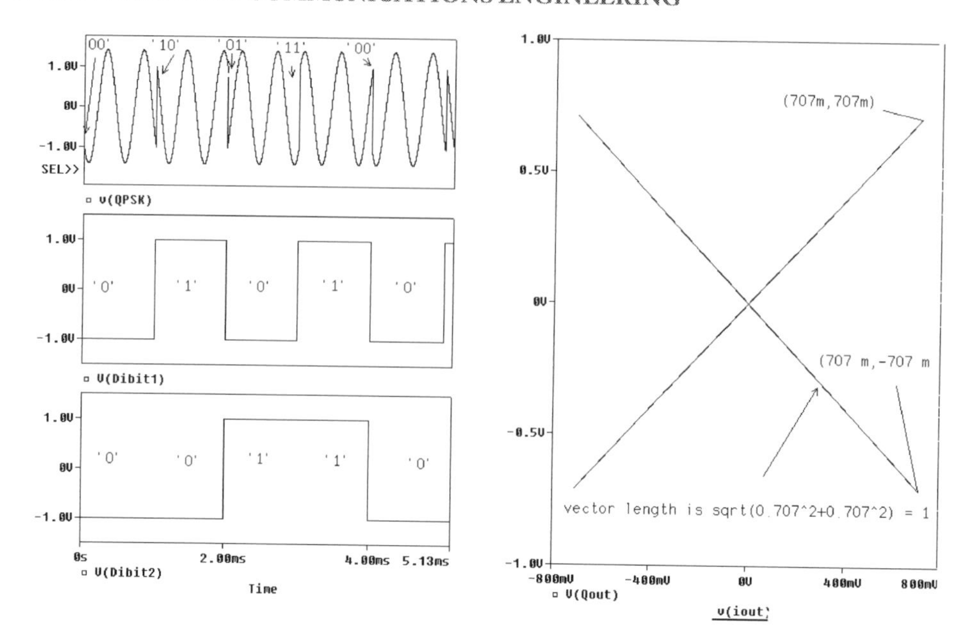

FIGURE 5.3: Bit-splitter signals.

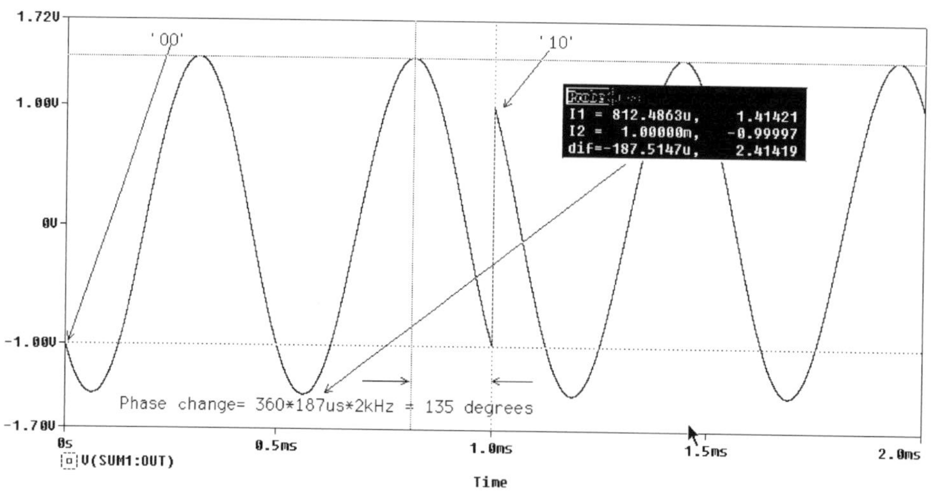

FIGURE 5.4: Measuring the phase change.

5.5.1 QPSK Modulation Using a Bit-Splitter

The QPSK modulator shown in Fig. 5.5 has the **ABM** parts in the QPSK transmitter replaced with actual circuitry. A hierarchical construction is used because of the system complexity.

Draw the blocks using the **Place hierarchical block** icon with the names as suggested and place inside each block the various functions. The "I" balanced modulator comprises two

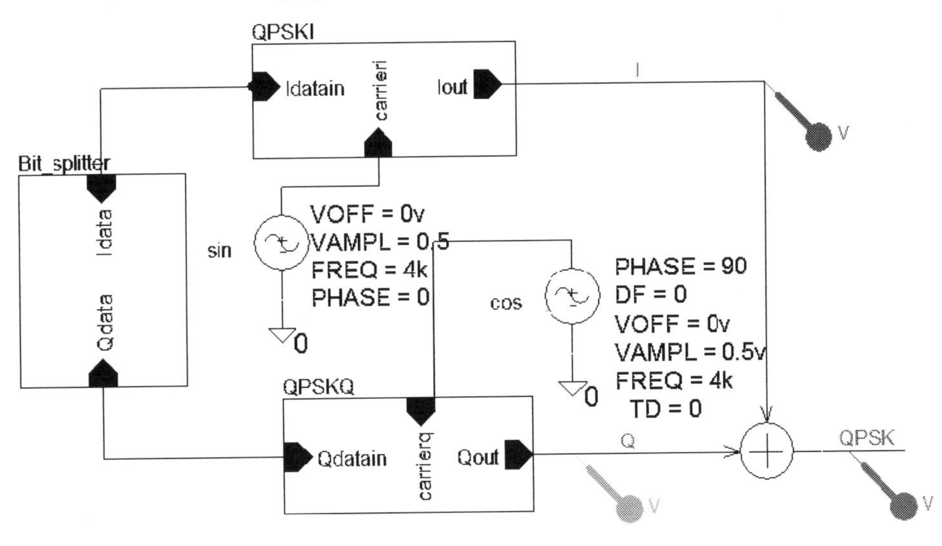

FIGURE 5.5: QPSK modulator.

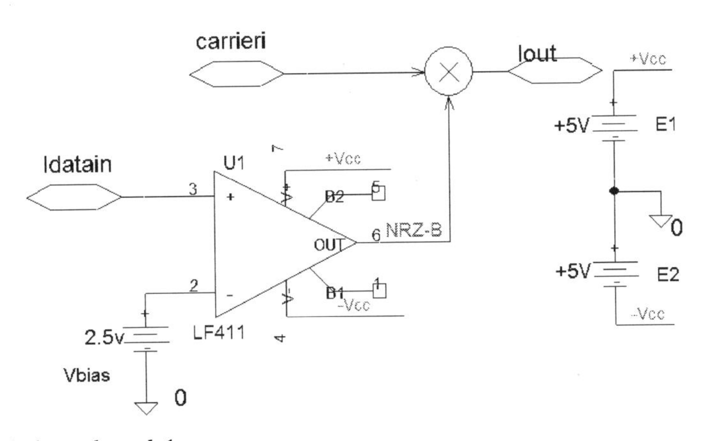

FIGURE 5.6: I-balanced modulator.

transformers with pairs of diodes to switch the transmission path as used in the previous balanced modulator, but here we use a simple **ABM MULT** part shown in Fig. 5.6. We need the path switching function to be in NRZ-B form, i.e., magnitude of ± 5 V.

Repeat this schematic for the "Q" BM but with different interface names. To delineate circuit functions, place a box by selecting the box from the right toolbar menu and draw it around the bit-splitter as shown in Fig. 5.7. This helps to identify complete sub-sections of a complex circuit. **Lclick** the box line and select the line type.

The data "110010111" is entered into a **STIM1** part using "relative time" with the plus symbol. For absolute time values, type in the time values but without the plus sign. This

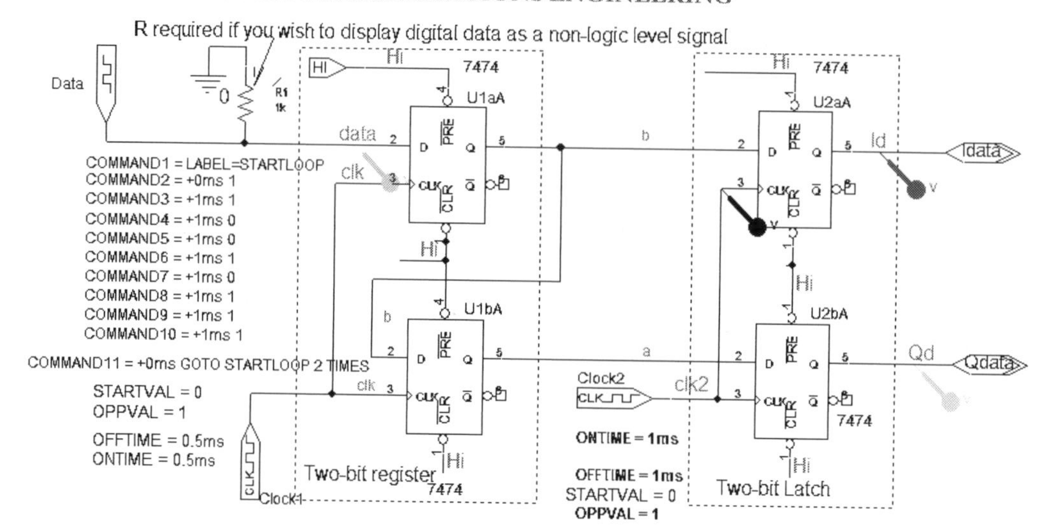

FIGURE 5.7: Bit-splitter.

sequence repeats two times using the STARTLOOP and GOTO STARTLOOP functions. Changing the command to GOTO STARTLOOP –1 TIMES produces an infinitely repeating sequence. The 2-bit register clock signal parameters are as shown in the schematic. The sine carrier uses the **VSIN** part with **VAMPL** = 0.5 V and **FREQ** = 10 kHz. The cosine carrier parameters are the same except **Phase** = 90°. The outputs from each modulator are then linearly summed using the **SUM** part. Set **Run to time** to 20ms, and **Maximum step size** to 1 μs. Select the **Simulation setting** menu and the **Options** tab. In the **Category** box select **Gate-level Simulation**. Set **Timing Mode** to **Typical** and the **Initialize all flip flops** to All **0**. Failing to do this will result in the flip-flop being in an undetermined state i.e., **All X** shows up in the plot as two red lines. Simulate with **F11** key to produce the signals as shown in Fig. 5.8.

I and *Q* signals in Fig. 5.9 are shown "stretched" in time compared to the original data signal.

Table 5.2 shows the relationship between the data, and "I" and "Q" signals.

TABLE 5.2: The Quadrature Signals

DATA	1	1	0	0	1	0	1	1	1	0
I		1		0		1		1		1
Q			1			0		0		1

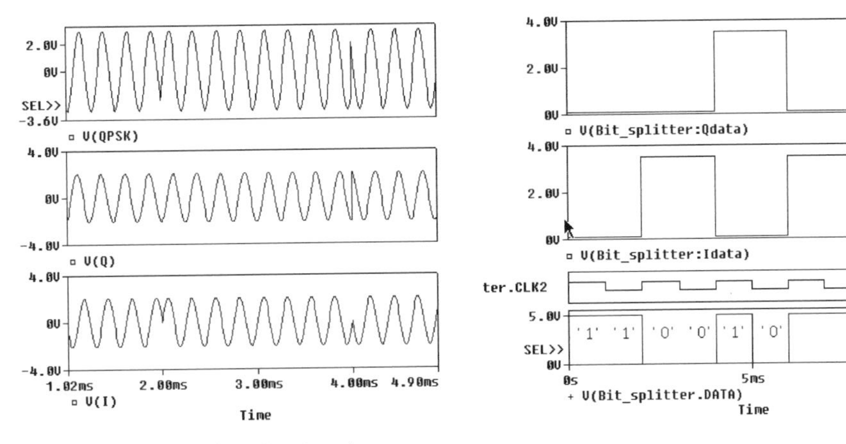

FIGURE 5.8: The NRZ bipolar signals.

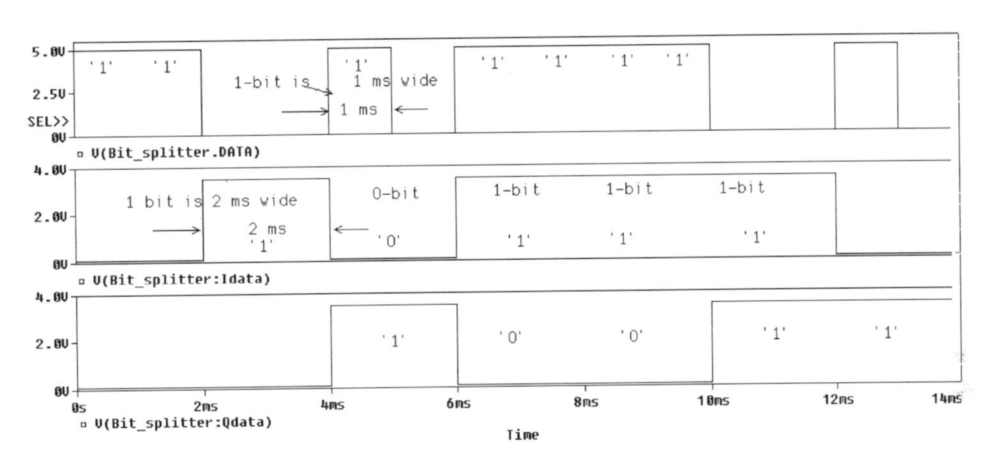

FIGURE 5.9: The I and Q data signals.

The NRZ-B data signals in the Q and I modulator blocks should be filtered by a root-raised cosine filter, but this is left as an exercise to investigate using the filters from Section 6.7.3.

5.5.2 QPSK Receiver

It is necessary to use quadrature cosine and sine oscillators in a 4-PSK (QPSK) receiver demodulator because of phase ambiguities in the received input data. This arrangement is called a vector modulator and can be used for up or down conversion. The receiver carrier local oscillator must be frequency and phase coherent, i.e., the same as the transmitter carrier and with low-pass filters at the output of each balanced modulator. To see why the filter is necessary, we apply Euler's formula to each sin and cosine carriers, in terms of complex exponentials, as

$$\sin \omega t = \frac{e^{j\omega t} - e^{-j\omega t}}{2j} \quad \text{and} \quad \cos \omega t = \frac{e^{j\omega t} + e^{-j\omega t}}{2}. \tag{5.17}$$

Multiplying these two sine carriers yields

$$\sin^2 \omega t = \frac{e^{j\omega t} - e^{-j\omega t}}{2j} x \frac{e^{j\omega t} - e^{-j\omega t}}{2j} = \frac{e^{j2\omega t} - 2e^{j0} + e^{-2j\omega t}}{-4} = 0.5 - 0.5 \cos 2\omega t. \tag{5.18}$$

Equation (5.19) shows a DC level and a signal whose frequency is twice the original carrier frequency. Similarly, multiplying a sine carrier by a cosine carrier yields

$$\sin \omega t \cos \omega t = \frac{e^{j\omega t} - e^{-j\omega t}}{2j} x \frac{e^{j\omega t} + e^{-j\omega t}}{2} = \frac{e^{j2\omega t} - e^{-j2\omega t}}{4j} = \sin 2\omega t. \tag{5.19}$$

The frequency is doubled but contains no DC. Fig. 5.10 shows a QPSK receiver with an input QPSK signal applied with a **VPWL_F_RE_FOREVER** part. This reads in a text file C:\Pspice\Circuits\signalsources\data\qpsksignal.txt saved from the output of the previous QPSK experiment. A combination of **ABM** parts and actual components is used in the design (see the exercise at the end of the chapter). The Zener diode and Schmitt trigger prevent the flip-flop from being overdriven and generating errors. Two clocks are required, one of which is twice the frequency of the other clock. The vdata "11001011" generator is included to compare the original and recovered data. Terminating any digital part output with an analog part, such as a resistance or capacitor, forces PSpice to insert an A/D device between the two components and displays the data with +5 V signal amplitude instead of being displayed as a logic level signal, and the FFT can then be applied.

Set **Output File Options/Print values in the output file** to 1ms, **Run to time** to 16ms, and **Maximum step size** to 1 µs and simulate with **F11**. QPSK signals in Fig. 5.11 show the unfiltered recovered data containing the high frequency, and below that is the filtered signal.

5.6 OFFSET QUADRATURE PHASE SHIFT KEYING

In QPSK, the carrier phase changes phase every $2T$ s and a 90° phase change will occur when $I(t)$ or $Q(t)$ changes sign but a phase shift of 180° occurs when both quadrature components change sign. Digital radio modems use a form of QPSK called offset QPSK (OQPSK) to overcome the difficulty introduced when the transmitter power amplifier goes momentarily to zero during a phase change through 180°. It is difficult to design power amplifiers with a linear response down to zero power output. Filtering a QPSK signal reduces the spectral side lobes producing a waveform that no longer has a constant envelope where a 180 degree phase shift shift in the carrier causes the envelope to go through zero briefly producing undesirable interfering frequency side lobes. Power fluctuations are minimized in OQPSK since

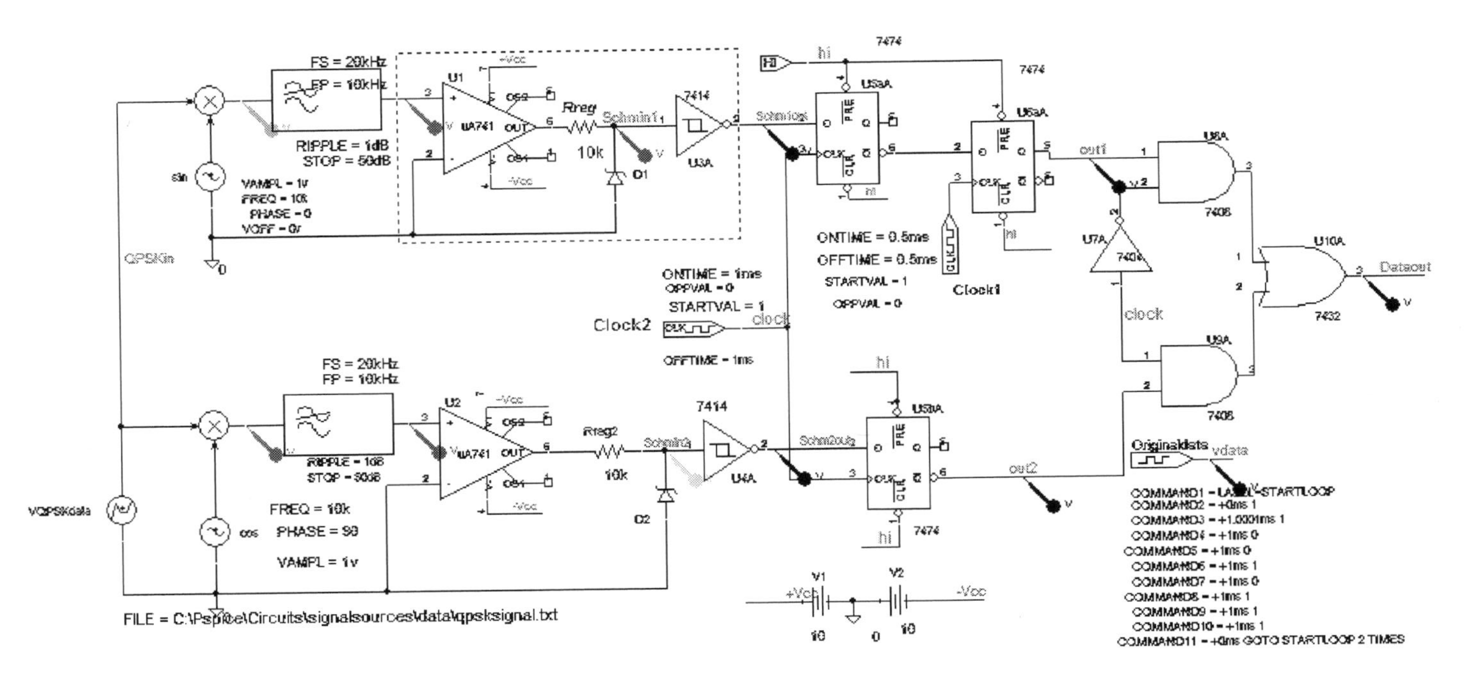

FIGURE 5.10: QPSK receiver.

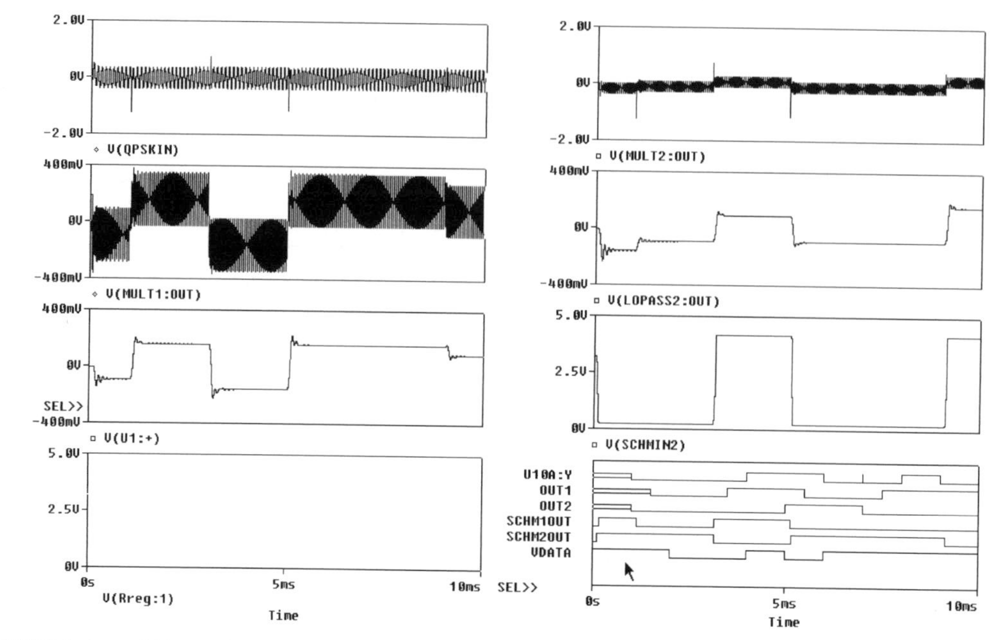

FIGURE 5.11: The unfiltered and filtered modulator signals.

the phase never actually changes by 180° so we modify the previous QPSK schematic to produce OQPSK. Fig. 5.12 shows an offset block to delay the "Q" bit stream with respect to the "I" bit stream by half a bit interval.

Again, remember that filters should be placed after the bit-splitter but we leave them out to see the actual processing more clearly. Replace the diode-balanced modulator with **ABM MULT** parts and the 741 opamp with an ideal operational amplifier called **OPAMP** in order to satisfy the evaluation version and enter the **DigClock** parameters for the offset block as

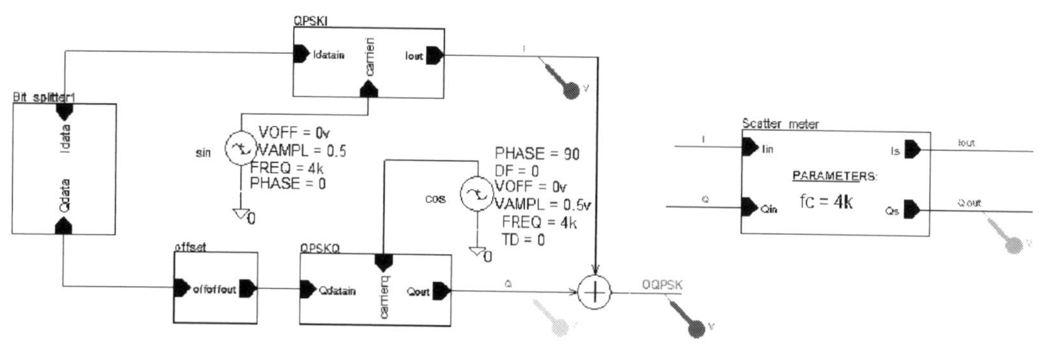

FIGURE 5.12: Offset QPSK block diagram.

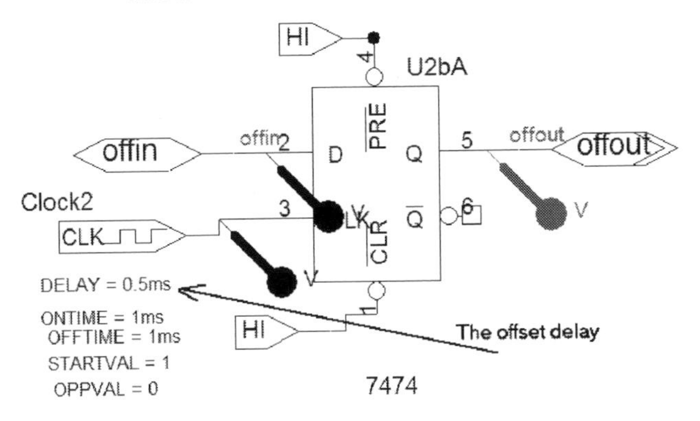

FIGURE 5.13: Inside of the offset block.

shown in Fig. 5.13. The offset is achieved by including another FF with the clock delayed by 0. 5ms.

Set **Output File Options/Print values in the output file** to 1ms, **Run to time** to 10ms, and **Maximum step size** to 10 μs. Simulate with **F11** key. The offset time signals for OQPSK are shown in Fig. 5.14. We can see that the phase change is now less than 180° but we can measure this using the technique explained previously.

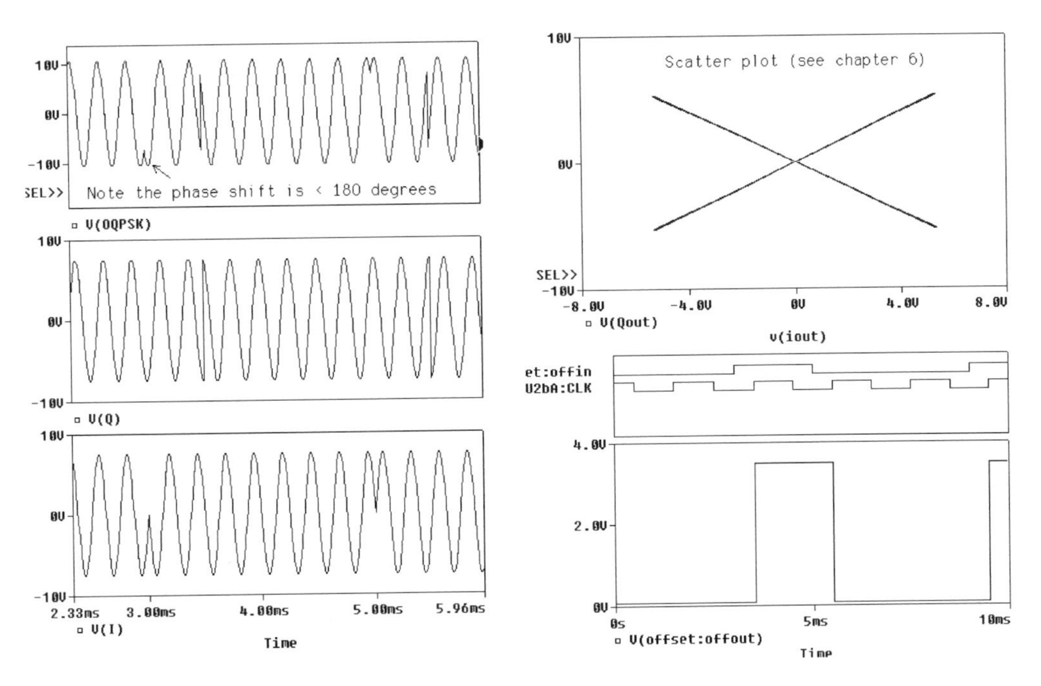

FIGURE 5.14: OQPSK waveforms.

5.7 GAUSSIAN MINIMUM SHIFT KEYING

Gaussian minimum shift keying (GMSK) is used in GSM modulation schemes in mobile radio systems. This produces a significant improvement over QPSK by modifying the shape of the transmitted pulses using a Gaussian filter prior to transmission. MSK uses a sine-shaped pulse, rather than rectangular pulse, and has a linear phase change limited to $\pm\pi/2$ over a bit interval. A linear phase change produces a power spectral density (PSD) with a larger main lobe and smaller side lobes and produces improved adjacent-channel selectivity compared to other systems. A Gaussian distribution is expressed as

$$g(t) = \frac{1}{2T}\left\{1 - \cos\left(\frac{2\pi t}{T}\right)\right\}. \tag{5.20}$$

The schematic in Fig. 5.15 investigates the effect of a Gaussian filter (window size Tms) on a square wave. The window width is defined in a **PARAM** part with $T = 1$ms.

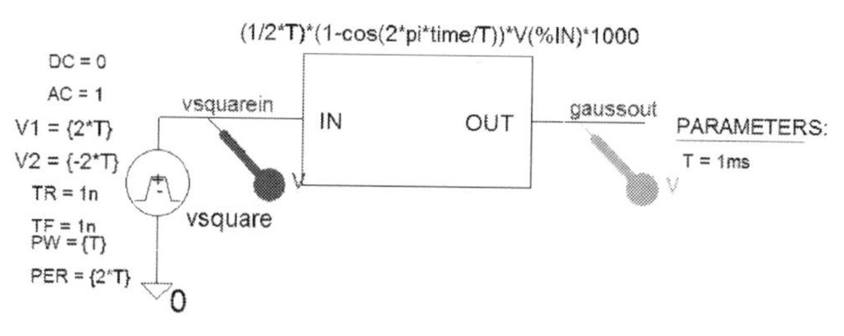

FIGURE 5.15: Gaussian filtering.

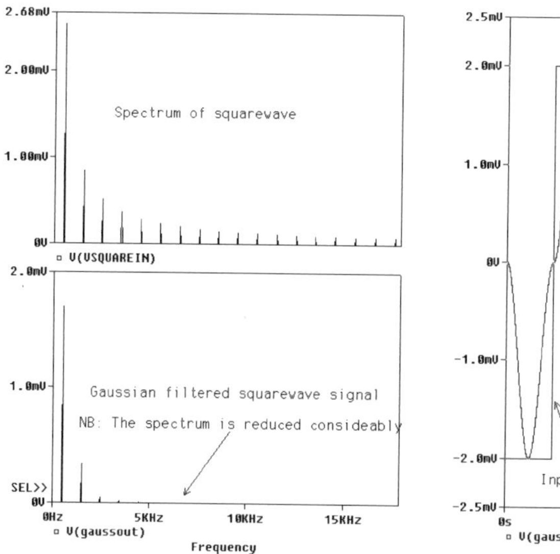

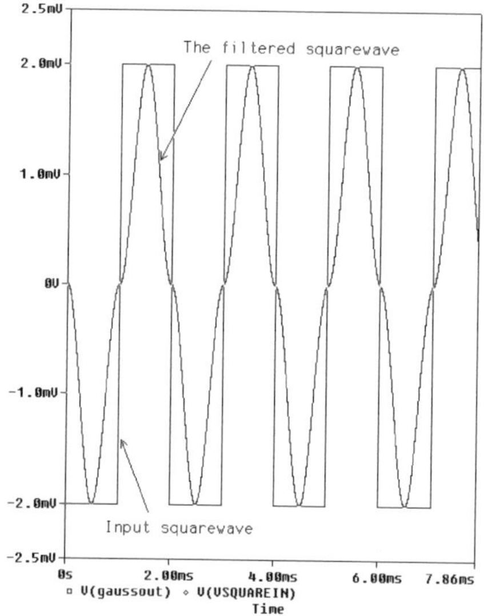

FIGURE 5.16: Input and output signals.

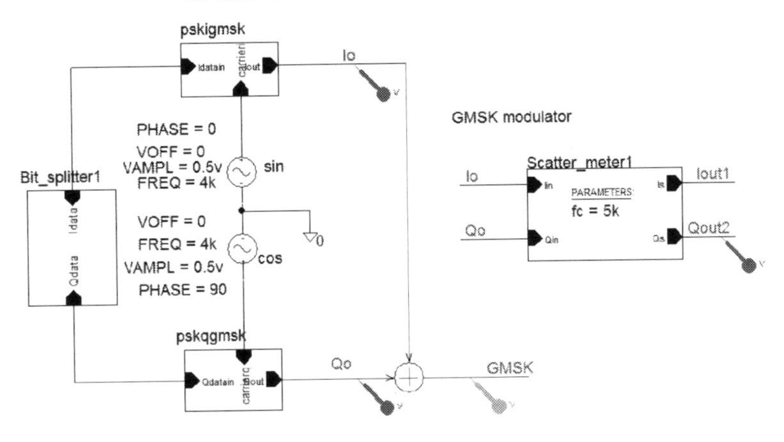

FIGURE 5.17: GMSK production.

Set **Output File Options/Print values in the output file** to 1ms, **Run to time** to 40ms, **Maximum step size** to 1 μs, and simulate with **F11** key. Compare the two time signals in Fig. 5.16.

The spectrum for the two signals shows the side-lobe reduction for the Gaussian-shaped signal leading to much less intersymbol interference (ISI). The QPSK system considered previously is redesigned to produce a basic GMSK system as shown in Fig. 5.17.

The Gaussian filter is the main difference between QPSK and GMSK. The balanced modulator is achieved with an **ABM** multiplier as shown in Fig. 5.18.

Set **Output File Options/Print values in the output file** to 10ms, **Run to time** to 40ms, **Maximum step size** to 0.1 μs. Simulate by pressing the **F11** key to produce the GMSK signals shown in Fig. 5.19.

5.8 EIGHT-PSK

Extending 4-PSK to eight carrier phase changes results in 8-PSK where the data is formed into 3-bit groupings called tribits and with the bits arranged in a gray code, as shown in

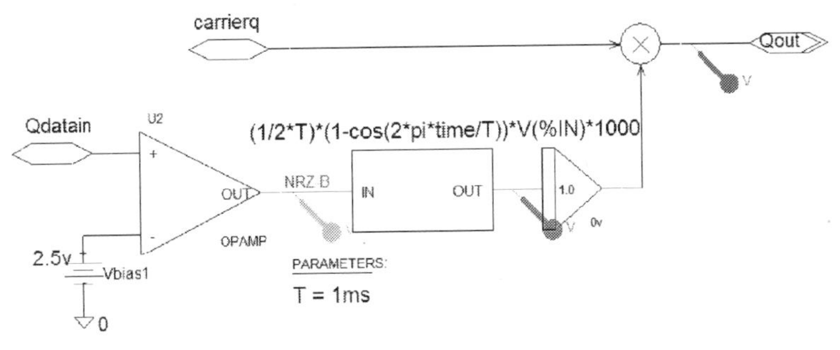

FIGURE 5.18: The Q BM using a **MULT** part.

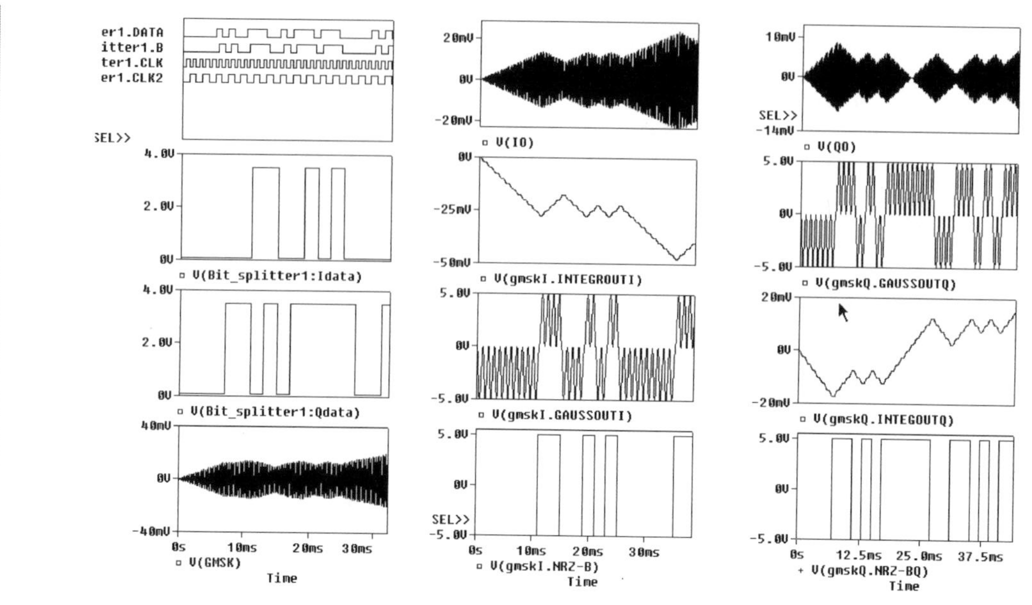

FIGURE 5.19: GMSK waveforms.

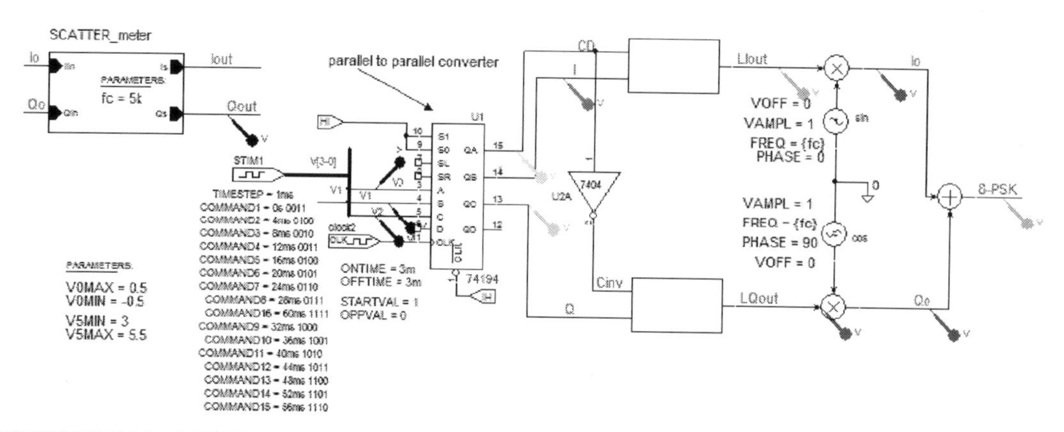

FIGURE 5.20: 8-PSK generation.

Table 5.3. This code arrangement leads to an improved bit error rate (BER) performance, as only one bit changes from one tribit to the next. In natural or binary code, there may be as much as a 3-bit difference in any data change, e.g., changing from 3 to 4 is a change 011 to 100.

Fig. 5.20 shows the **ABM** components replaced with actual IC parts. For example, use a 74194 IC as a parallel-to-parallel converter to create tribit signals. **ABM1** parts define the levels for each tribit signal as per Table 5.4.

TABLE 5.3: Binary and Gray Code Arrangements

N	ORDER	BIT CHANGE	GRAY CODE	BIT CHANGE	θ
1	111	1	100	1	315
2	000	3	000	1	0
3	001	1	001	1	45
4	010	2	011	1	90
5	011	1	010	1	135
6	100	3	110	1	225
7	101	1	111	1	180
8	110	2	101	1	270
9	111	1	100	1	315

```
EXP1   If(((V(CD)>v0min)&(V(CD)<v0max))&((V(I)>v0min)&(V(I)<v0max)),-0.54,0)
EXP2   +If(((V(CD)>v5min)& (V(CD)<v5max))&((V(I)>v0min)&(V(I)<v0max)),-1.307,0)
EXP3   +If(((V(CD)>v0min)& (V(CD)<v0max))&((V(I)>v5min)&(V(I)<v5max)),0.541,0)
EXP4   +If(((V(CD)>v5min)& (V(CD)<v5max))&((V(I)>v5min)&(V(I)<v5max)),1.307,0)
```

FIGURE 5.21: *M*-ary converter 1.

```
EXP1   If(((V(Cinv)>v0min)&(V(Cinv)<v0max))&((V(Q)>v0min)&(V(Q)<v0max)),-0.541,0)
EXP2   +If(((V(Cinv)>v5min)& (V(Cinv)<v5max))&((V(Q)>v0min)&(V(Q)<v0max)),-1.307,0)
EXP3   +If(((V(Cinv)>v0min)& (V(Cinv)<v0max))&((V(Q)>v5min)&(V(Q)<v5max)),0.541,0)
EXP4   +If(((V(Cinv)>v5min)& (V(Cinv)<v5max))&((V(Q)>v5min)&(V(Q)<v5max)),1.307,0)
```

FIGURE 5.22: *M*-ary converter 2.

Set **Output File Options/Print values in the output file** to 40ms, **Run to time** to 100ms, and **Maximum step size** to 1ms, and simulate with the **F11** key. The test data is a 4-bit digital stimulus generator called **STIM4** and outputs four parallel signals, the structure of which is shown in Fig. 5.20. A bus system for connecting signals is drawn using the icon (shortcut **B**). Enter the bus names using the net alias icon.

The level-shifter logic in the **ABM** parts is shown in Figs. 5.21 and 5.22.

The 8-PSK waveforms are shown in Fig. 5.23 together with a scatter diagram obtained using the meter investigated in Chapter 6.

TABLE 5.4: Tribit Levels

I	C	LEVEL
0	0	−0.541
0	1	−1.307
1	0	0.541
1	1	1.307
Q	**CN**	**LEVEL**
0	1	−1.307
0	0	−0.541
1	1	1.307
1	0	0.541

Plotting orthogonally, the sampled Q output v(Qout) with the x-axis changed to the sampled I output, v(Iout), results in the scatter diagram showing the spatial relationship between the tribits in Fig. 5.23 (discussed in Chapter 6). It really is a **VECTOR** diagram, but PSpice will not allow you to plot points without connecting them together, so the spokes stay unfortunately. The **PROBE** output might have a nonsquare aspect ratio (depends on your computer screen), so you might have to resize the plot by grabbing a side of the plot and resizing to a square aspect ratio. We see in the vector diagram that one of the vectors is given as

$$V_{8-\mathrm{PSK}}(\theta) = 1.307 \sin\theta + 0.541 \cos\theta. \tag{5.21}$$

So for an angle of pi/8 $= 22.5^\circ$ this results in a magnitude of 1. Note that this angle is with respect to the real axis but the angle between constellation points is $360/8 = 45^\circ$.

The various digital signals shown in Fig. 5.24 should be investigated.

5.9 QUADRATURE AMPLITUDE MODULATION

Quadrature amplitude modulation (QAM) is a method of modulating a carrier using a combination of carrier amplitudes and phase shifts. 16-QAM is used in the 9600 bps (2400 baud × 4 bits) modem, with each point in the constellation representing 4 bits of data, and is also used in digital microwave links. There is a better separation between constellation points when compared to PSK, which leads to better noise immunity and hence better bit error rate (BER)

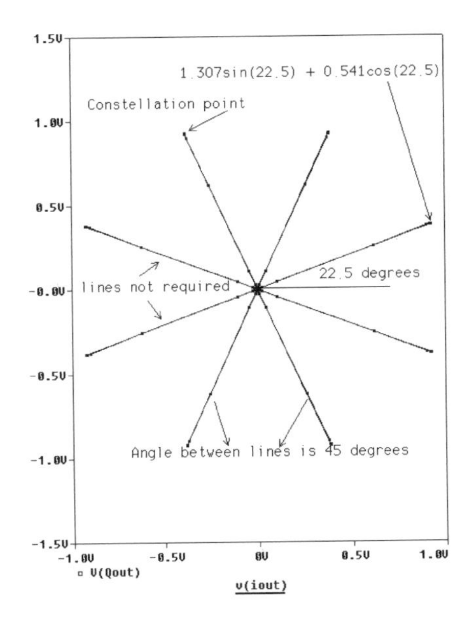

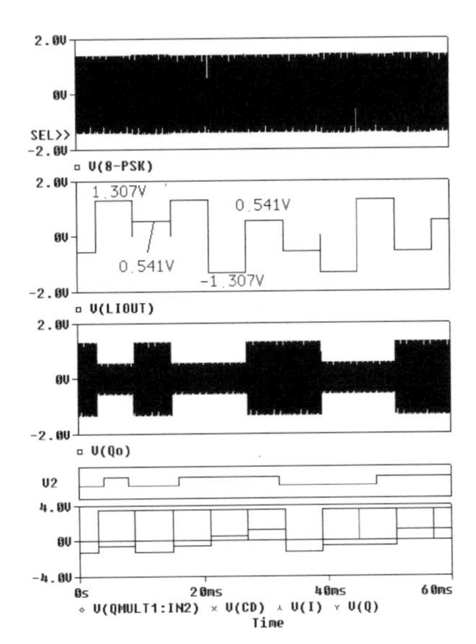

FIGURE 5.23: Scatter plot and 8-PSK waveforms.

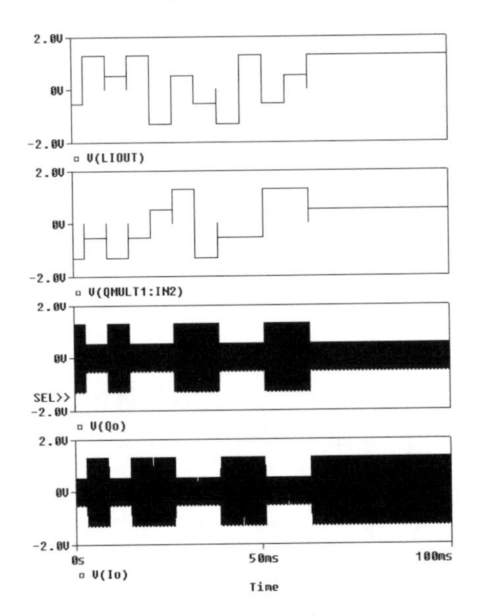

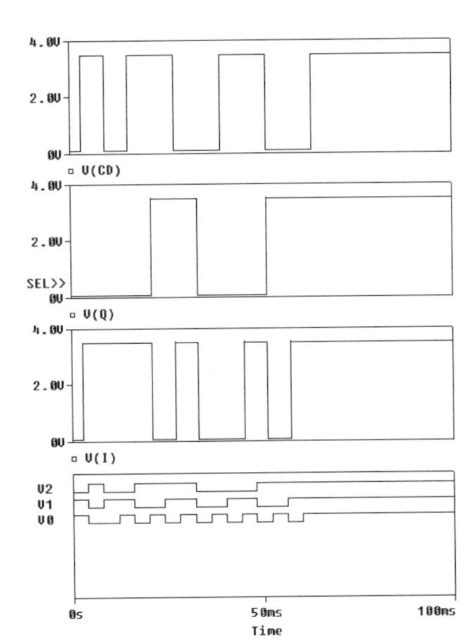

FIGURE 5.24: 8-PSK digital signals.

performance. We compare PSK and QAM by assuming that each system has the same power, and the same number of points, M. 16-PSK has points that are spaced uniformly around the circle, with each point close to its neighboring points. Increasing the circle radius means higher transmitting power, which is the same for QAM. Transmitting 1 W of power results in the amplitude of the PSK circle being 1 (A^2 = power).

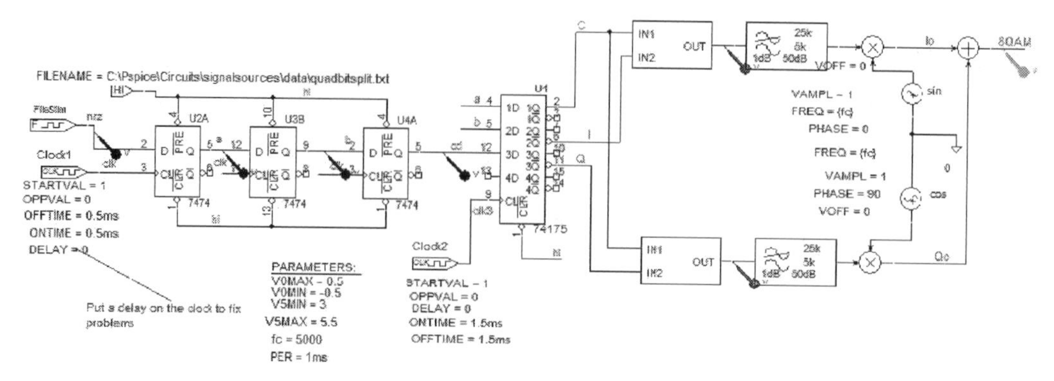

FIGURE 5.25: 8-QAM modulator.

5.9.1 Eight-QAM

An 8-QPSK signal is created using the modulator displayed in Figure 5.25 by splitting an NRZ signal into three signals a, b, c, and applying them to quadruple D-type flip-flop with clear (74175) IC. The three outputs are fed to two **ABM1** parts to create multi-level signals for application to the two modulators but unlike 8-PSK, we do not invert the CD signal into the lower ABM part. Filters are placed at the output of the level-shifter as shown but leave them out initially in order to obtain a clear scatter diagram as displayed in Fig. 5.23.

Filters are placed at the outputs of each level shifter as shown but you should leave them out initially in order to obtain a clear scatter diagram as displayed in Fig. 5.26.

5.9.2 Comparison of PSK and QAM

The vector amplitudes in QAM are not all equal. Some vectors having larger or smaller amplitudes when examined in the constellation. We may work out the spacing between nearest neighboring points in QAM, if each constellation point is assumed to have an equal probability of occurring. QAM constellation is more efficient in distributing (spacing) points for a fixed amount of power than PSK. This translates into a better overall performance in the presence of noise giving it a bigger advantage over PSK as M increases. PSK, however, has a constant-amplitude property making it less prone to interference in a channel with nonlinear characteristics, making it easier to demodulate and recover data bits. Consider the partial constellation diagram in Fig. 5.27. Here we compare a 16-PSK map to a 16-QAM map. An error might occur when a noise signal, A_n, has an amplitude that is equal to, or greater than, half the distance between constellation points $A_n \geq 0.5h_n$. Let us assume that the noise is such that $A_n = 0.5h_n$. In 16-PSK the angle between points is $360/16 = 22.5°$ so that the relationship between A and A_n is $\sin\theta/2 = \frac{h/2}{A} \Rightarrow h/2 = A\sin\theta/2$.

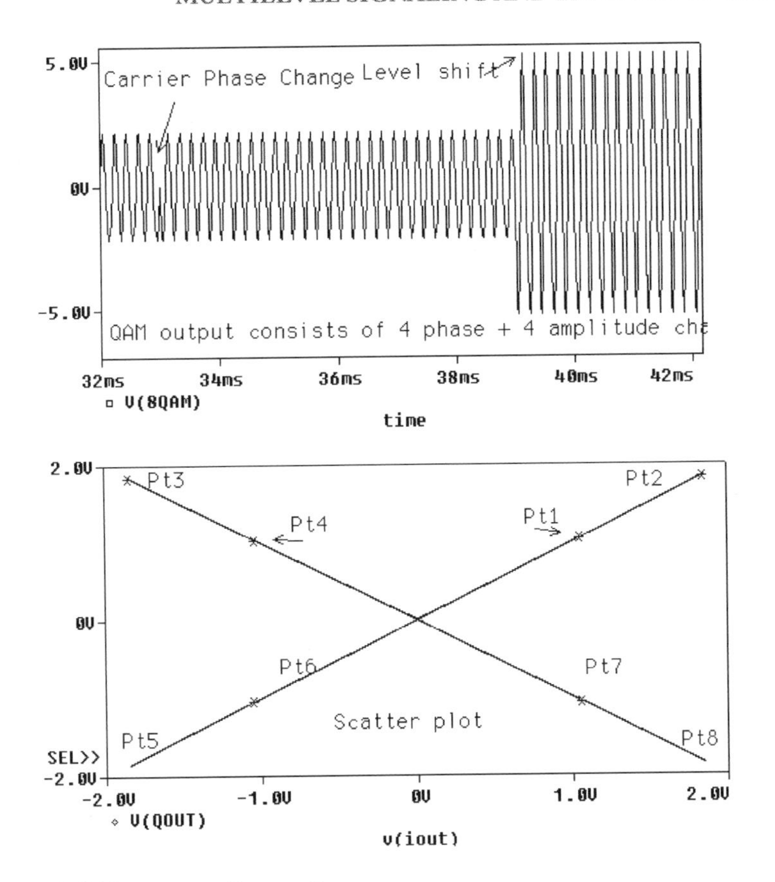

FIGURE 5.26: 8-QAM output + Scatter diagram.

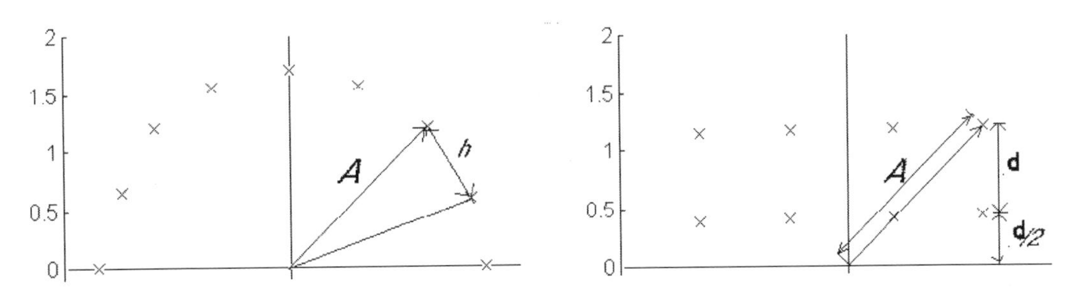

FIGURE 5.27: 16-PSK versus 16-QAM.

Half the angle between points in a 16-PSK constellation is $\theta/2 = (22.5/2)^\circ$, and the noise amplitude is $A_n = 0.5h = A\sin(22.5/2) = 0.195$. Consider the spatial relationships between noise and signal vectors in a 16-QAM plot. Here again an error might occur when a noise signal A_n has an amplitude that is equal to or greater than half the amplitude of the wanted

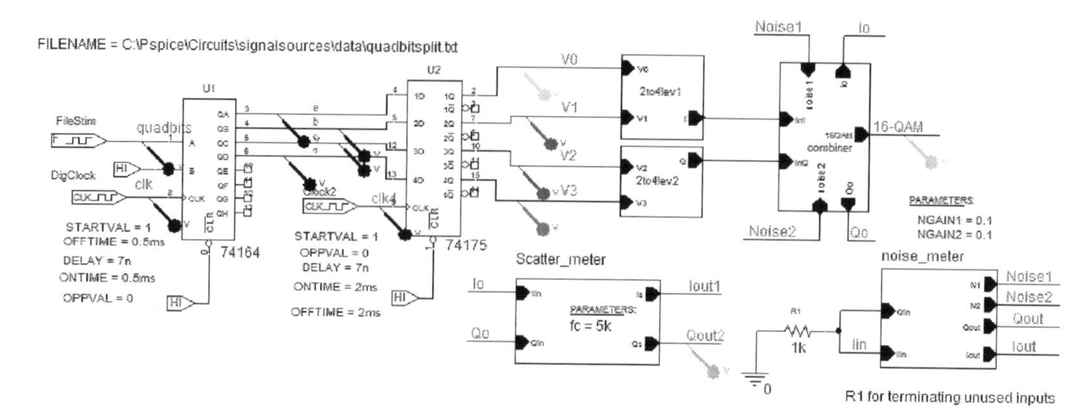

FIGURE 5.28: 16-QAM modulator.

signal constellation spacing, i.e., $A_n = 0.5d$; then we can write that A and A_n are

$$A^2 = 2(d + d/2)^2 = 2\left(3\frac{d}{2}\right)^2 = 2(3A_n)^2 = 18A_n^2 \Rightarrow \frac{A}{A_n} = 18 \Rightarrow \frac{A_n}{A} = 0.2357.$$

The ratio between the signal and noise vectors for the two systems in dB is

$$20\log\left(\frac{A_{n-\text{QAM}}}{A_{n-\text{PSK}}}\right) = 20\log\left(\frac{0.2357}{0.195}\right) = 1.58\ \text{dB}.$$

For equal transmitted powers, 16-QAM has better noise immunity compared to 16-PSK. The noise may increase to $0.2357A$ for 16-QAM on peaks, whereas for 16-PSK, using the same power on max signals, the noise peaks are allowed to be $0.195A$. This implies better noise performance for the 16-QAM ($0.235A/0.195A$), which is 20.5% or 1.6 dB. From the two constellation diagrams we can see that the average power for 16-QAM is less than for 16-PSK.

5.9.3 Sixteen-Quadrature Amplitude Modulation

16-QAM expands the previous circuit. Fig. 5.28 shows a 16-QAM modulator using **ABM** and actual parts. The shift registers split the input signal into quad bits, which are then processed using two, two- to four-level, **ABM** parts to generate *in-phase* and *quadrature* signals that modulate orthogonal carriers to occupy a single channel centered at the 2 kHz carrier. Each symbol is a specific combination of signal amplitude and phase.

The input data from the text file shown in Fig. 5.29 covers 4-bit permutations 0000 to 1111 and enables us to plot the **VECTOR**/constellation diagram. This file was read in using a **FileStim** generator part attached to a quadbit splitter consisting of a 74164 register, and a 74175 IC to latch the four signals from the register. The serial register clock is 1 kHz and the

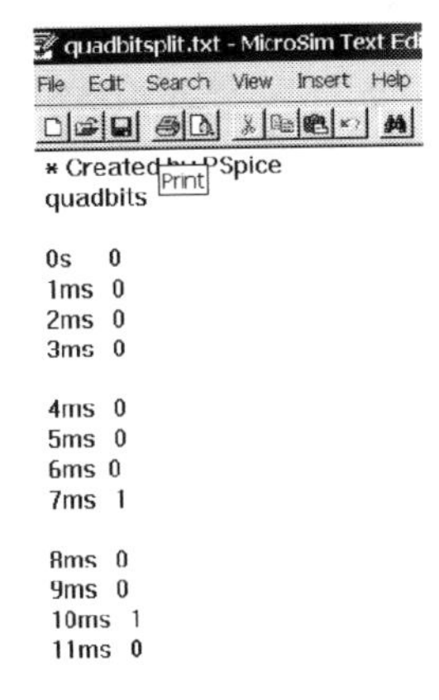

FIGURE 5.29: The input data file.

latch register clock is a quarter of that at 250 Hz and stretches the data signal to four times the input data length.

Warnings, as shown in Fig. 5.30, might occur after simulation. You are invited to plot to see what is causing the condition to merit warnings in the first place. This problem was "fixed" by adding a 7 ns delay to the input clock.

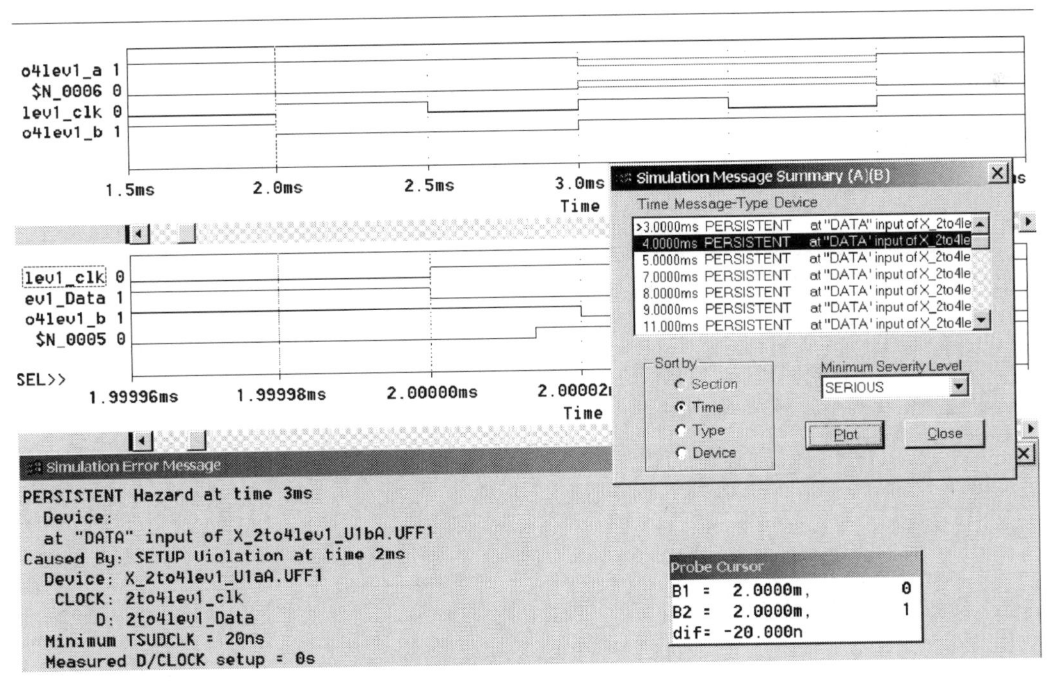

FIGURE 5.30: Warnings fixed by placing delays on the clocks.

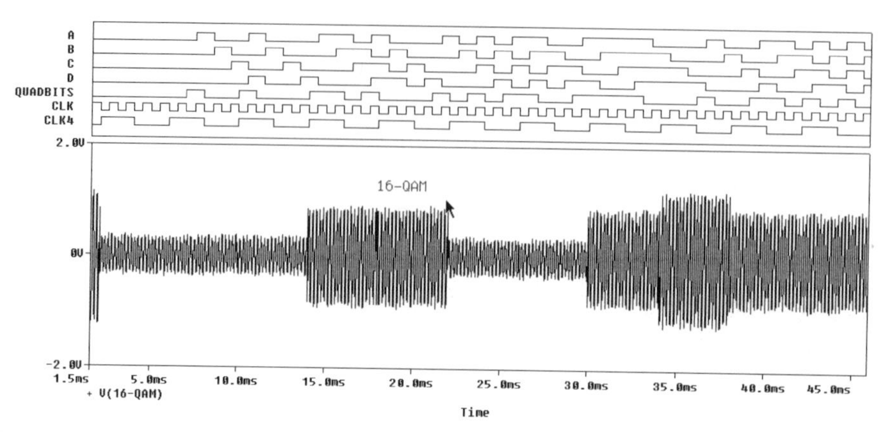

FIGURE 5.31: Input data signals.

The digital signals are shown in Fig. 5.31.

The four quad bit signals are shown in Fig. 5.32.

5.9.4 Two-to-Four Level Conversion

I and *Q* level signals are shown in Fig. 5.33. The two- to four-level converter is implemented using an **ABM2/1** part containing IF, THEN, ELSE statements. The 2-to-4 lev2 block in Figs. 5.33 and 5.34 shows the bit-splitter producing *Q* and *I* components. The bit-splitter contains two clocks with clock2 running at twice the speed of clock1. The four signals V0–V3 are applied to **ABM2/1** parts to achieve the required 16-QAM voltage levels (thanks to Lee Tobin for sorting out my fuzzy attempts at two- to four-level conversion):

```
If(((V( V0)>v0min)& (V(V0)<v0max)) &((V( V1)>v0min)& (V(V1)<v0max)),–0.22,0)
```

What this means that if V0 is approx 0, and V1 is approx 0, then the output is –0.22.

We are keeping this design as simple as possible but it should be remembered that a filter should be added at the output of the level shifter. Apply two orthogonal carriers to the **MULT** parts as in Fig. 5.35 to form a vector modulator.

The multilevel 16-QAM signal is shown in Fig. 5.36.

Select the **FFT** icon to examine the QAM frequency spectrum in Fig. 5.37. To plot the constellation diagram, change the *x*-axis from time to v(xa) and the *y*-axis is v(ya), as shown below. A detailed explanation of the scatter meter is examined in Chapter 6.

5.10 CLOCK EXTRACTION

A common requirement for baseband receivers is to extract a clock signal from the incoming decoded signal. There are several methods for extracting a carrier from the signal, and one

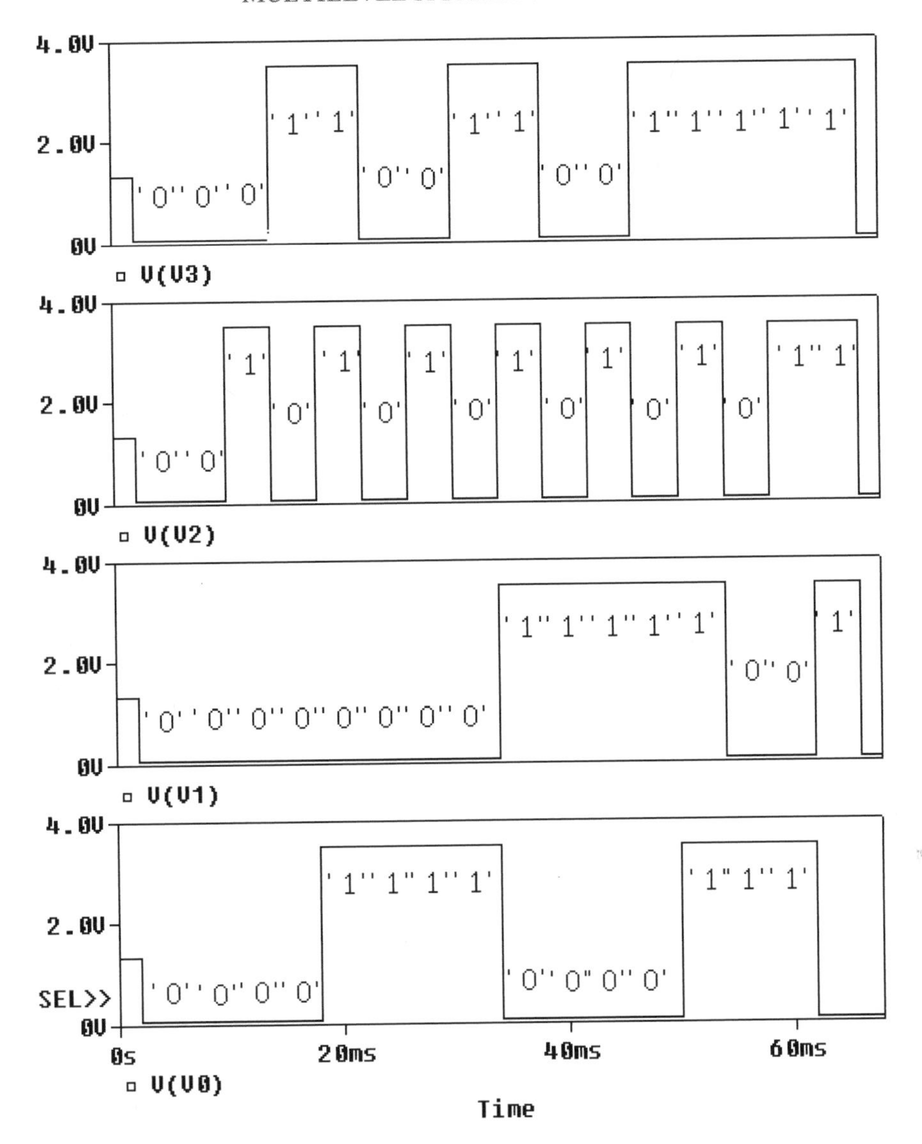

FIGURE 5.32: Quadbit signals.

method uses a PLL as shown in Fig. 5.38. Here a 565 IC acts as the voltage-controlled oscillator (VCO) (select the part, **Rclick**, and mirror horizontally to flip the IC through 180°), and a 7408 AND gate acts as the phase detector. The input signal is provided by a 4 V **VPULSE** generator with a **PRF** = 1 kHz and **PER** = 1ms. A noise signal is also applied using the **VPWL_F_RE_FOREVER** generator part and is added to the data using a **SUM** part. In the **Transient Analysis** menu place a tick on **Skip initial transient solution** and simulate.

Fig. 5.39 shows how the PLL locks onto the data after approximately 2ms.

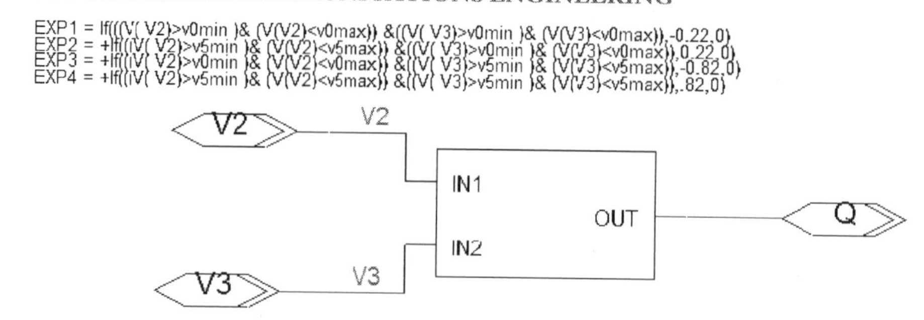

FIGURE 5.33: The *Q*-channel two- to four-level converter.

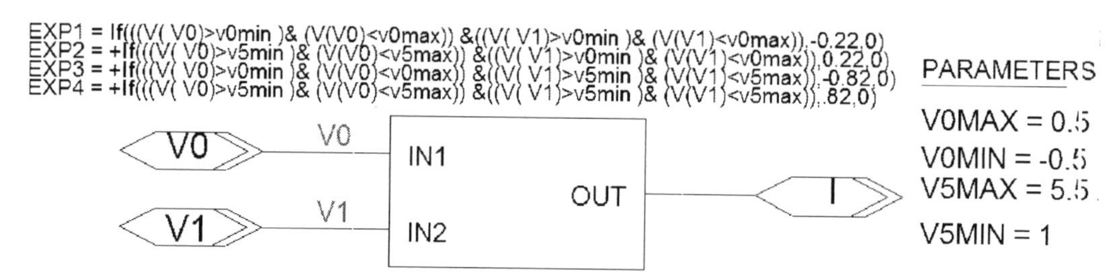

FIGURE 5.34: *I*-channel two to four-level converter.

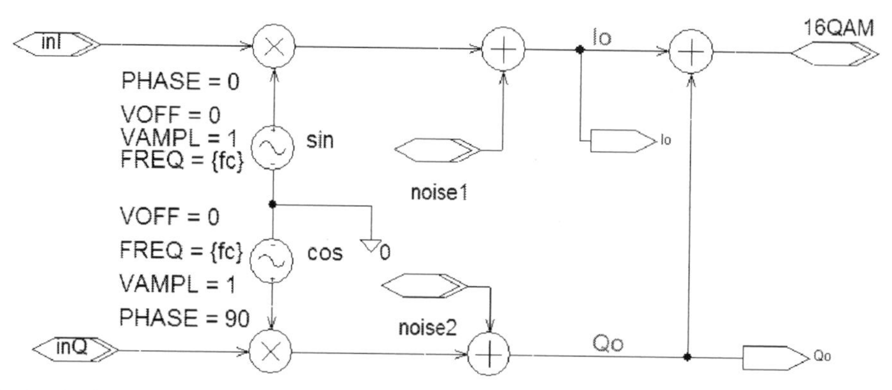

FIGURE 5.35: Summing the quadrature and in-phase components.

The LPF integrates the gate output and causes the VCO to change its frequency to match the input frequency. Investigate locking action by varying the amount of noise, or reduce the data signal level. However, warnings may occur if you increase the noise level past the AND gate maximum input level.

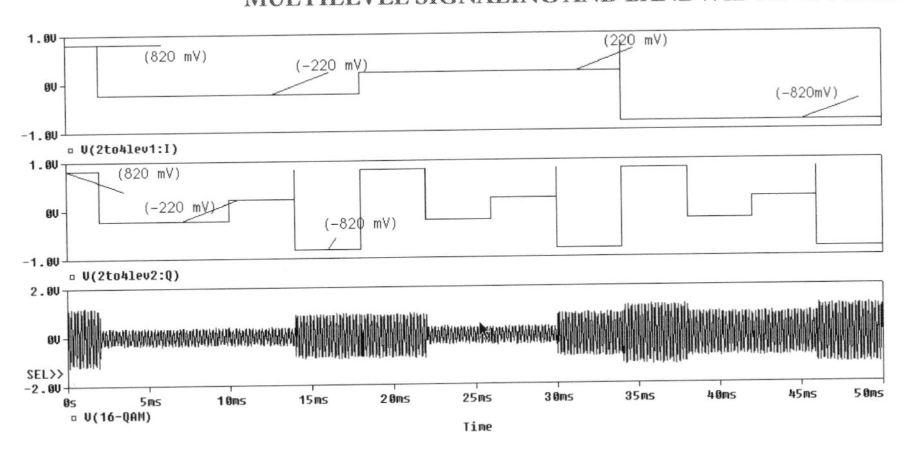

FIGURE 5.36: A 16-QAM signal.

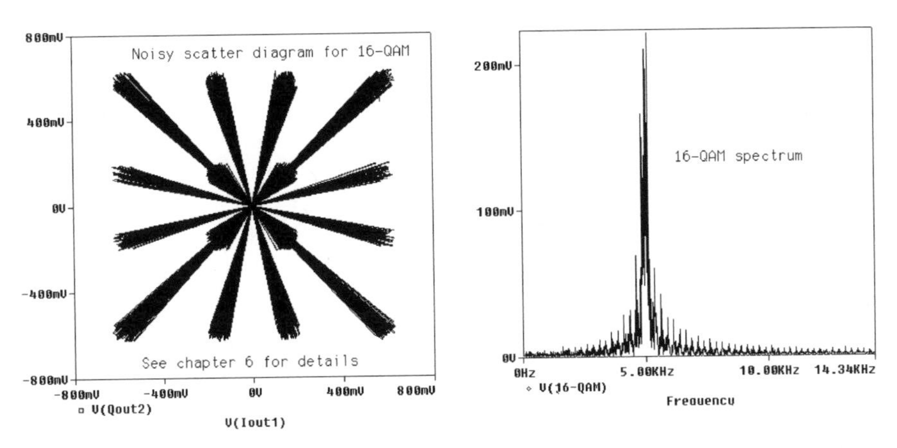

FIGURE 5.37: 16-QAM spectrum.

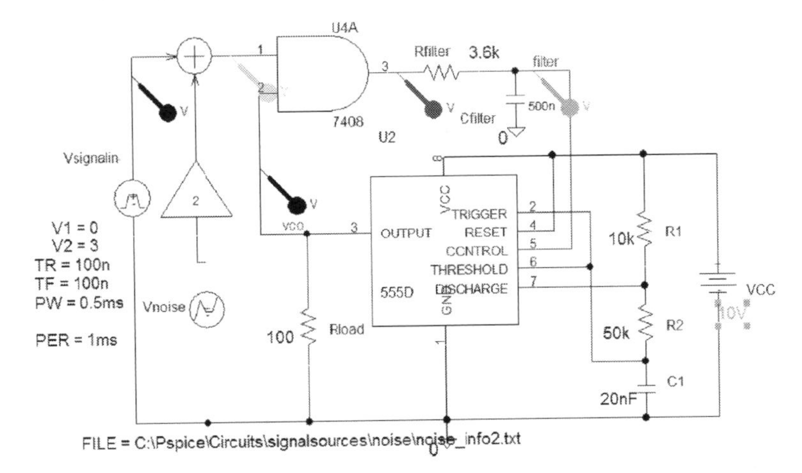

FIGURE 5.38: A PLL using the 555 IC.

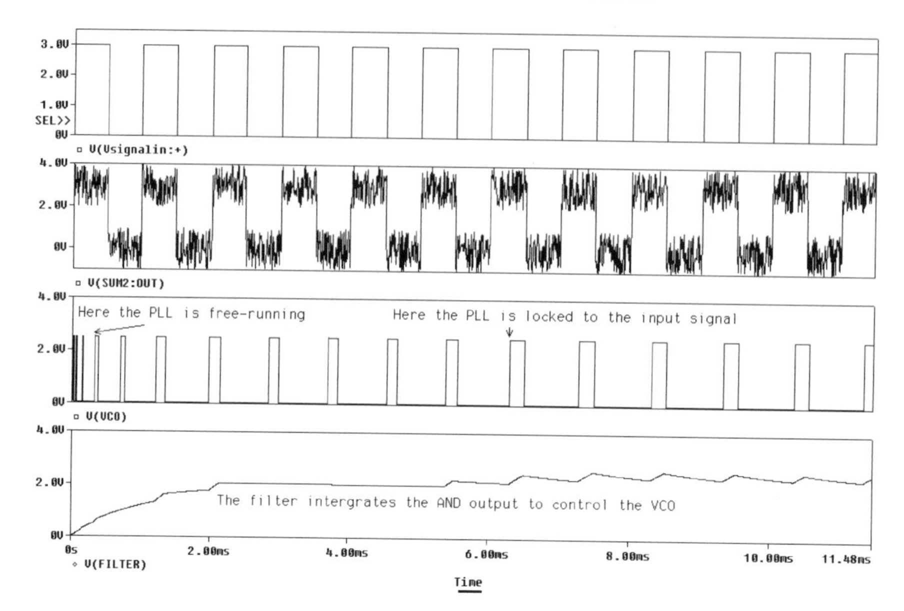

FIGURE 5.39: PLL time waveforms.

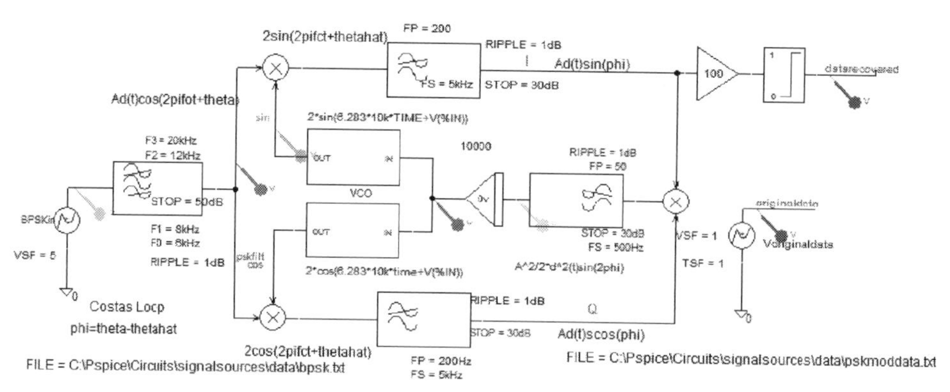

FIGURE 5.40: Costa loop.

5.11 COSTAS LOOP

The Costas loop in Fig. 5.40 is a special PLL that extracts a coherent carrier from a BPSK signal which has no carrier signal component in the spectrum, but uses the sidebands to produce a coherent carrier.

The binary-phase shifted (BPSK) 10 kHz carrier signal is applied as a text file called PSKin.txt to the circuit using a **VPWL_F_RE_FOREVER** part. This file was created by pasting the copied output variable from a **PROBE** result after simulating PSK into the Notepad program. Load the PSK signal and apply to the loop via a bandpass **ABM** filter in order to

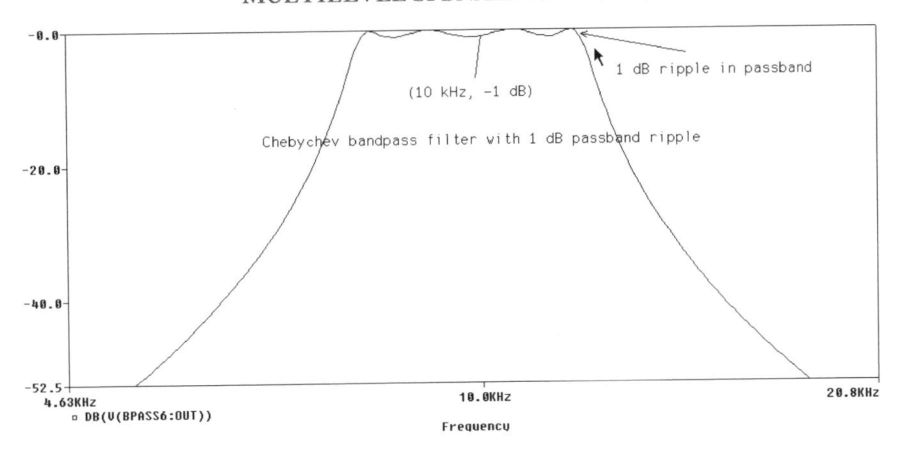

FIGURE 5.41: Bandpass frequency response.

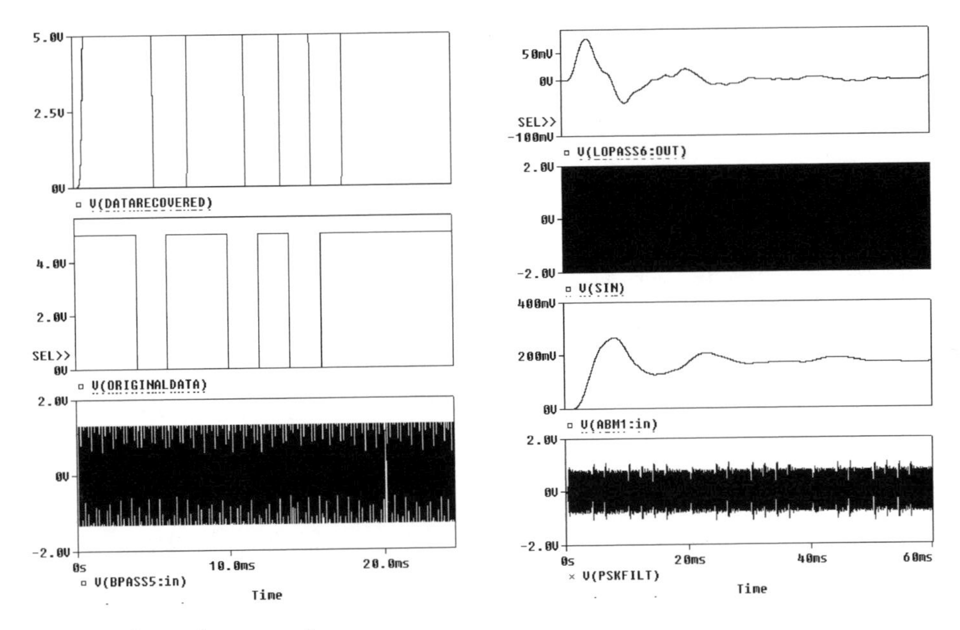

FIGURE 5.42: Costas loop waveforms.

limit the frequency content. The frequency response for this filter is shown in Fig. 5.41. The VCO **ABM** equation is 2*sin(2*pi*10k*TIME+V(%IN)), where the voltage V(%IN) from the integrator forces the VCO to 10 kHz.

The quadrature cosine oscillator uses another **ABM** part, but the equation is now 2*cos(6–283*10k*TIME+V(%IN)). The Costas loop signals are shown in Fig. 5.42.

The voltage-controlled oscillator (VCO) frequency is controlled from a multiplier whose inputs are the quadrature outputs from two multipliers. The multiplier output voltage is pro-

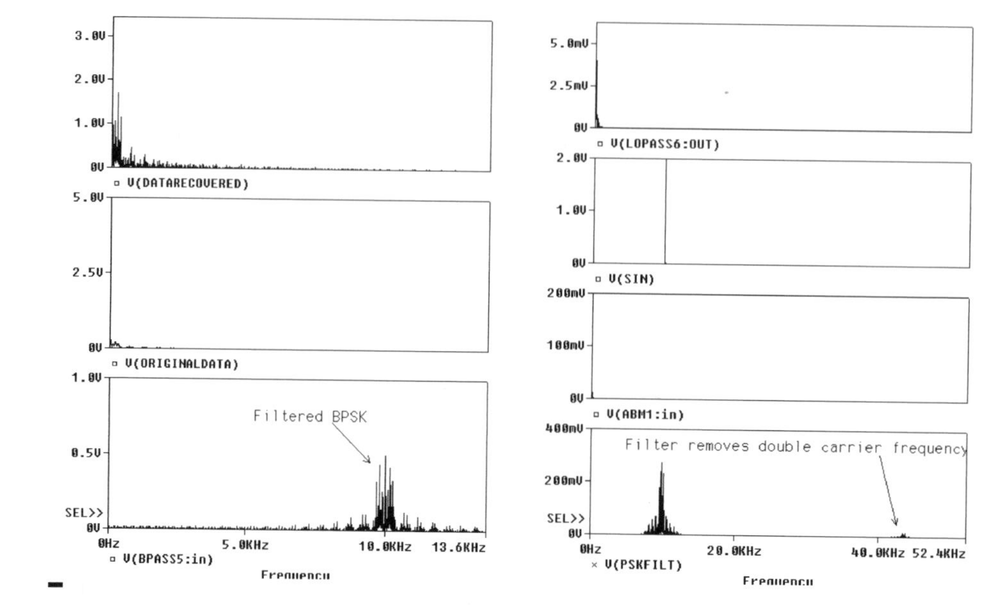

FIGURE 5.43: Costas loop spectra.

portional to the phase error of the local oscillator. The low-pass filters have a cut-off frequency that is determined by the data rate (in this case the fundamental frequency of the data signal is 200 Hz). The outputs from each multiplier are

$$y_I(t) = d(t)A\sin(\phi) \tag{5.22}$$

$$y_Q(t) = d(t)A\cos(\phi). \tag{5.23}$$

The output from the third multiplier (a phase doubler) is

$$z(t) = y_I(t) * y_Q(t) = d(t)A\sin(\phi) * d(t)A\cos(\phi) = \frac{A^2 d^2(t)}{2}\sin(2\phi). \tag{5.24}$$

The third filter removes spurious components that could change the VCO frequency. Investigate if the loop local oscillator tracks a slightly different carrier frequency; i.e., will the loop force the VCO to change its frequency? The Costas loop spectra are displayed in Fig. 5.43.

The original PSK modulating data was included for comparison purposes.

5.12 EXERCISES

1. Modify the QPSK transmitter discussed previously but replace the **MULT ABM** part by the four-quadrant multiplier AD633 and the LPF by a second-order active filter.

FIGURE 5.44: GMSK.3.

2. Investigate the effect of using a raised cosine filter on the final output of the QPSK spectrum.
3. Realize the block diagram with a suitable design for a GMSK.3 system shown in Fig. 5.44. The 8-kbit bipolar data signal is filtered by a Gaussian filter whose bandwidth is 2.4 kHz, and applied to a DC-coupled FM modulator with a modulation index of 0.5. The ratio of 2.4 kHz to 8 kbps is 0.3, hence the name.
4. Add noise to the data in the GMSK.3 circuit and investigate, using the eye diagram, for different noise amplitudes.
5. Fig. 5.45 shows another phase-shifting technique for producing a 90° phase shift between two signals, but over a band of frequencies. Investigate the schematic and measure the phase between Vin and the voltage measured using the differential markers.
6. Investigate the QPSK modem in Fig. 5.46. Place a low-pass filter on the demodulated output and compare the recovered signal to the original dibits.
7. A PRBS is applied as a test data source for a phase shift keying (PSK) modulation circuit shown in Fig. 5.47. The PRBS is level-shifted making the PRBS symmetrical about zero and the output is fed to a simple multiplier circuit.

   The PSK output is shown in Fig. 5.48.
8. The tribit splitter in Fig. 5.49 splits the data using three D-type flip-flops and a 74175 quadruple D-type flip-flop latches the data. The test data at C:\Pspice\Circuits\signalsources\data\bitsplit.txt is applied with a **FileStim** part. The 8-PSK waveforms are illustrated in Fig. 5.50.

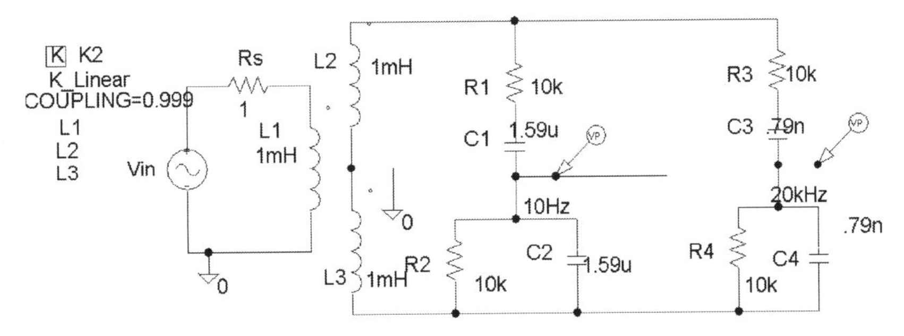

FIGURE 5.45: 90° phase shifter.

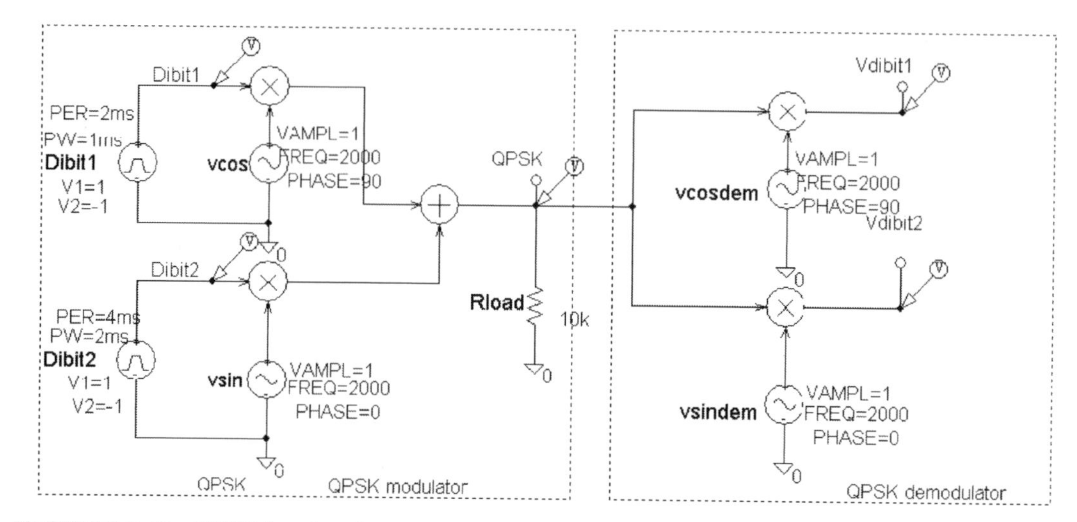

FIGURE 5.46: QPSK "modem."

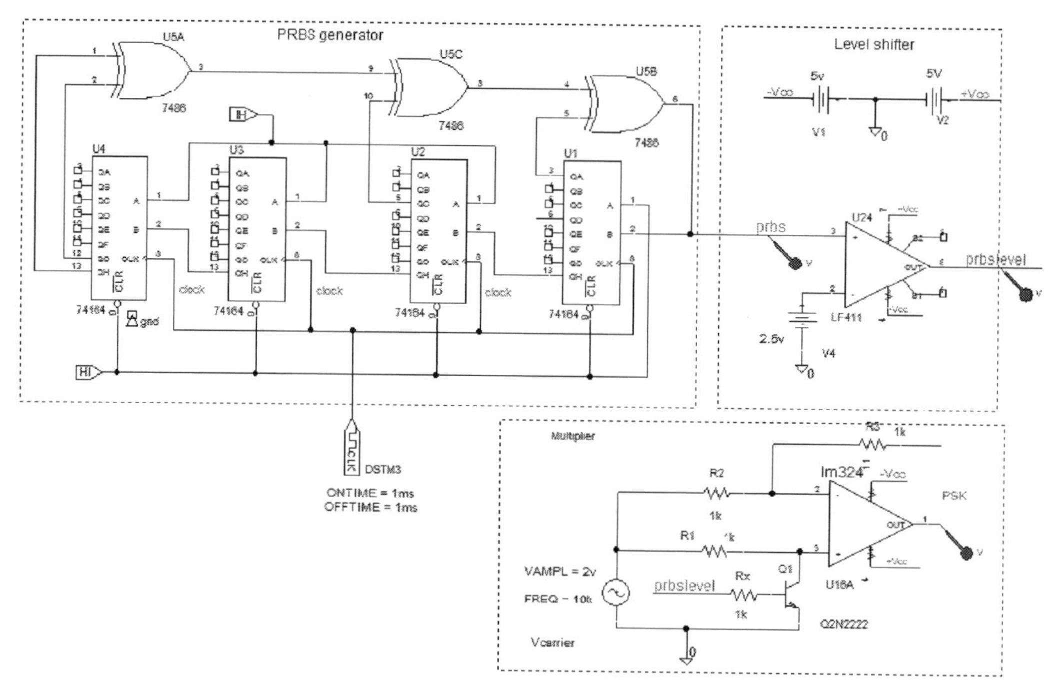

FIGURE 5.47: PSK modulator.

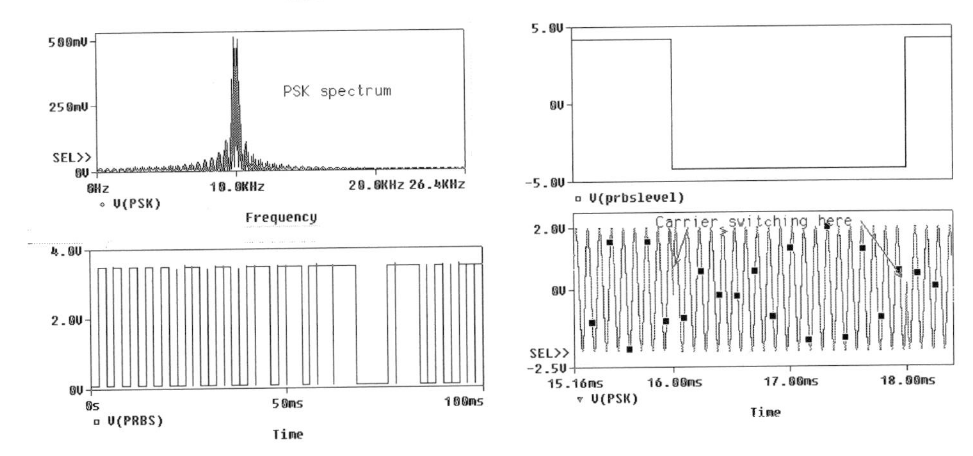

FIGURE 5.48: PSK modulator.

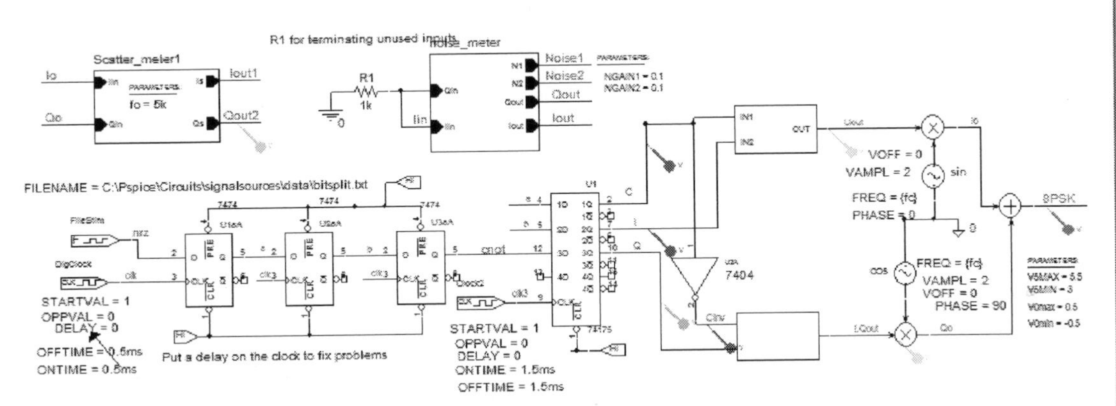

FIGURE 5.49: Tribit splitter and 8-PSK.

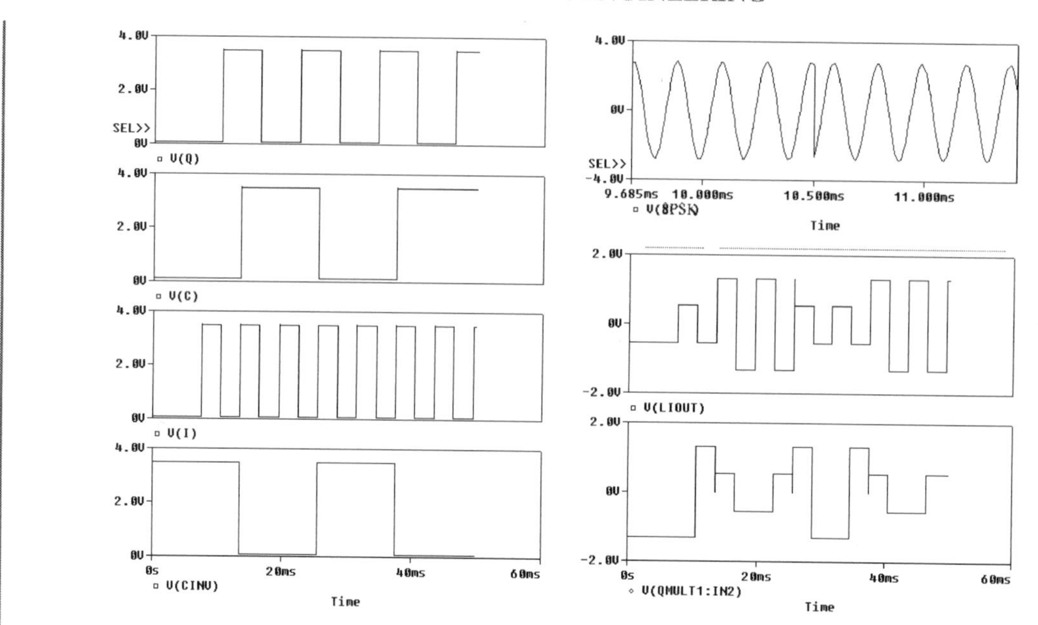

FIGURE 5.50: The 8-PSK waveforms are illustrated in Fig. 5.50.

CHAPTER 6

System Performance and Test Instruments

6.1 NOISE GENERATOR

Noise corrupts signals at all points in a telecommunications system and we need to look at a noise generator for investigating how noise changes system performance. We need to look at a noise generator for adding noise to any system for an investigation into how noise changes the parameters. Fig. 6.1 shows our noise meter which can be added to any schematic when required. There are two noise files in ASCII format created in Matlab and contain in a subfile accessed by selecting the meter, **Rclicking** and selecting **Descend hierarchy**. We need two independent noise sources to add to the two quadrature lines. The generator has several output lines for different applications but the first application investigates how noise changes the scatter diagram and uses noise1 and noise2 outputs. The noise amplitude for noise1 and noise2 can be changed by changing **NGAIN** parameter in the **PARAM** part.

We will also use this meter with the eye diagram as discussed in the next section.

6.2 EYE DIAGRAM

The eye diagram is a useful diagnostic tool for measuring the degradation in received data signals. It gets its name from the "eye pattern" resembling a human eye when binary data is examined in a special way. The eye is formed by folding the digital time signal back on itself from the right-hand side of an oscilloscope display to the left-hand side a number of times. The folding in PSpice is achieved by changing the x-axis, from a linear function of time, to a saw-toothed signal generated using a **VPULSE** part. In a real eye measurement system, the x-axis is time not voltage but this is not a problem since we can calibrate the voltage axis as time. The number of symbols displayed is a function of the pulse width (**PW**). For example, making **PW** twice the data period will result in two symbols being displayed.

The eye diagram assesses the quality of the received data in a modem by observing on an oscilloscope the effects of pulse distortion, intersymbol interference (ISI), and noise. The

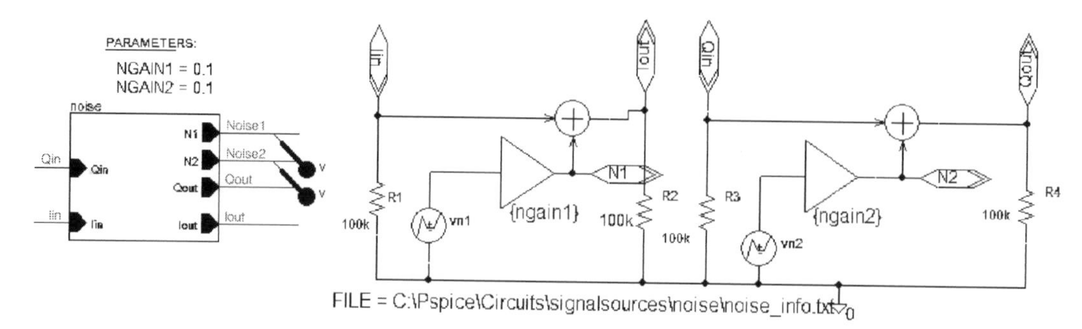

FIGURE 6.1: Noise generator/meter.

received signal is connected to the vertical channel and the system clock to the horizontal channel. At the sampling time, the received signal magnitude should be above or below the threshold voltage to ensure reliable detection of a binary 1 or 0. The eye opening size shows the voltage margin against added noise. The best position for the receiver voltage threshold and sampling instants is at the center of the eye opening. The noise generator simulates channel noise and makes the "eye" close. The pattern also shows if the equalization network is doing its job by showing the noise margin above which errors will occur in the receiver. Jitter can also be determined by measuring the deviations in time on the eye at the zero crossover point. The rate of eye closure is also a measure of timing inaccuracies, but is not simulated here.

6.2.1 The Eye Meter

The output from the *Q* or *I* output of an *M*-ary modulator is displayed using the eye meter in Fig. 6.2.

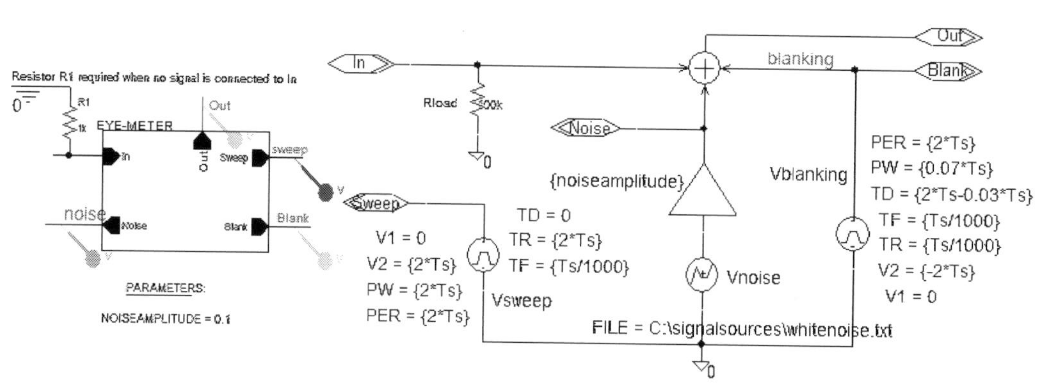

FIGURE 6.2: Eye meter.

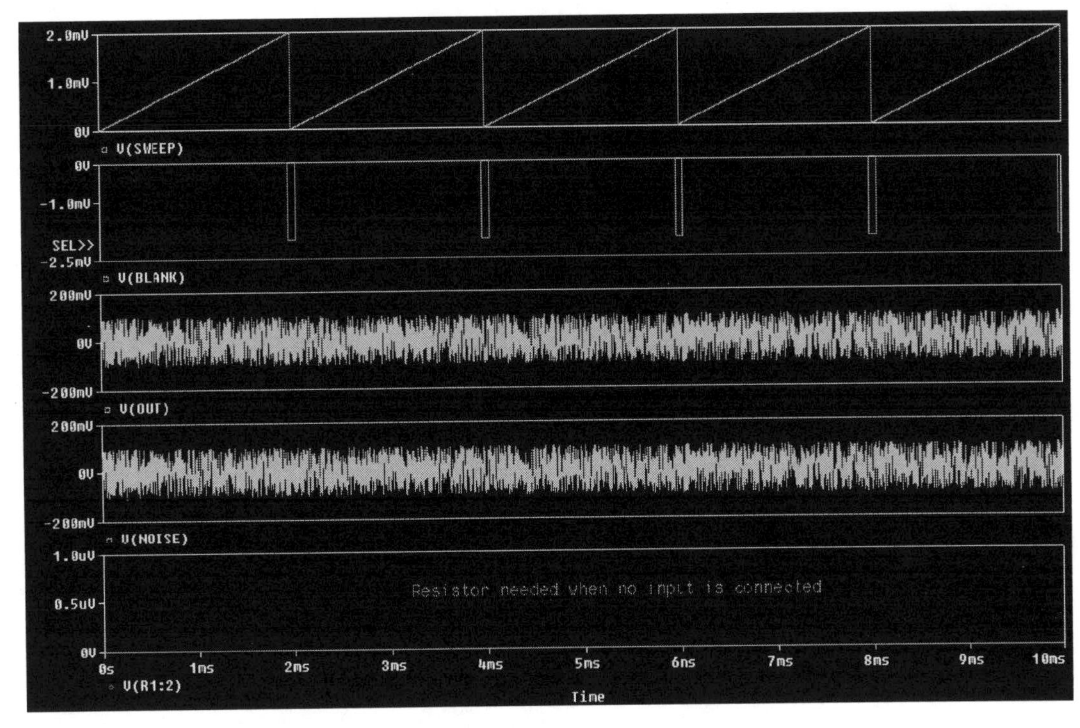

FIGURE 6.3: The eye diagram for a multilevel signal.

The blanking pulse is to stop the fly-back lines being visible (equivalent to z-modulation on a scope). Change the x-axis to the saw-tooth sweep signal whose period is twice the data rate period. To illustrate the effect on the signal by the channel (filter) bandwidth, we simulate with different bandwidths. The eye meter signals are shown in Fig. 6.3.

6.2.2 Eye Diagram Application

Fig. 6.4 shows a second-order active filter at the output of the PRBS circuit.

The filter mimics a limited-bandwidth channel and demonstrates how it changes the transmitted data signal characteristics. Noise is also added to the filter output to simulate the noise that is present in all systems. The noise amplitude is controlled by the **GAIN** part so that the eye may be investigated for different noise conditions. The time base sweep is provided by a **VPULSE** part with parameters as shown. The pulse rise time is set to the pulse width and so produces a saw-tooth signal rather than a pulse. It is important to set the pulse width and rise time to twice the data rate. Initially set the gain to zero (no noise). Press **F11** to produce the filtered PRBS signal. The output diagrams are shown in Fig. 6.5.

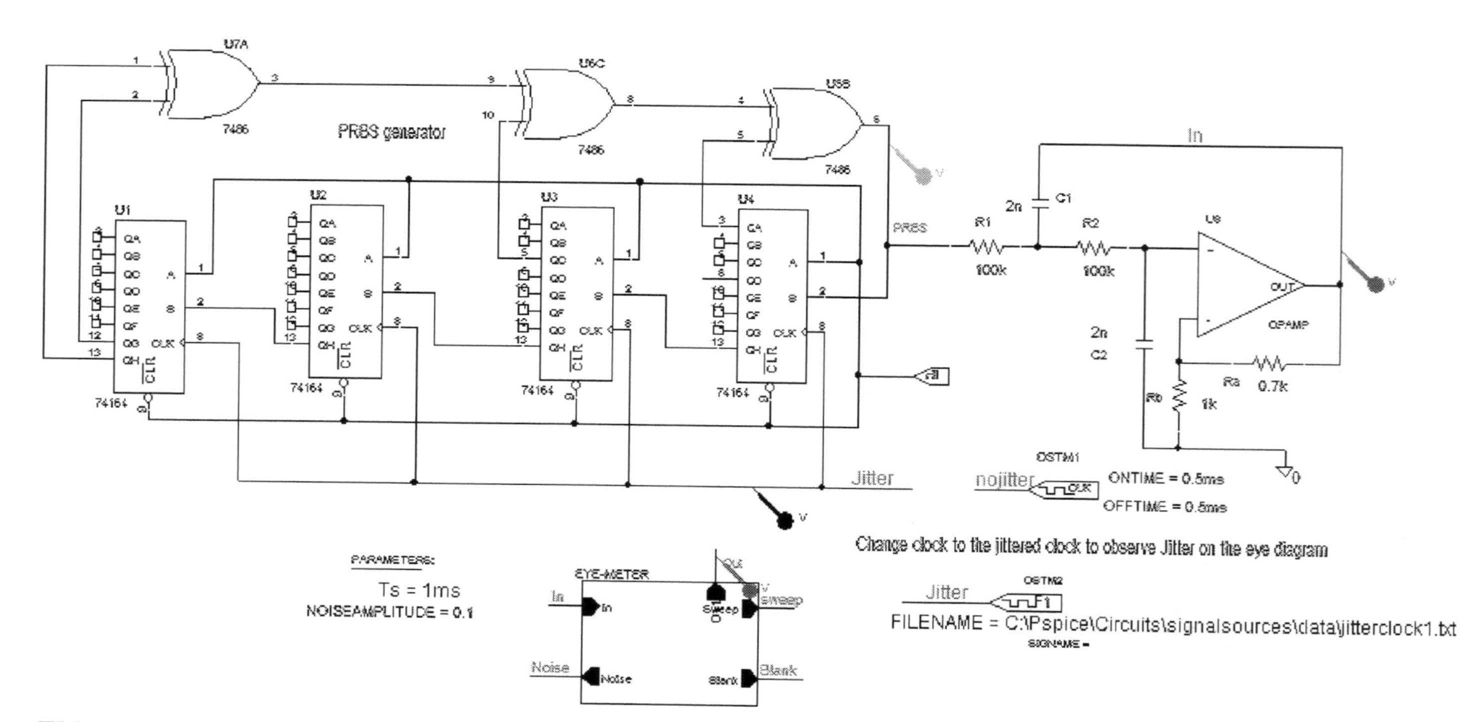

FIGURE 6.4: Applying the eye meter to a PRBS.

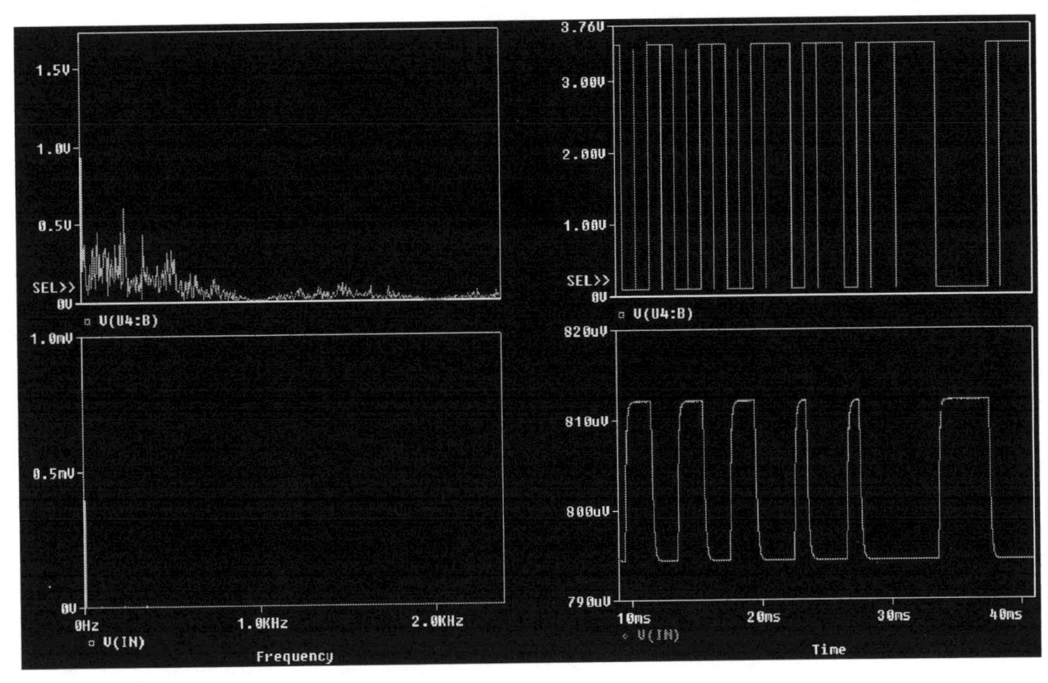

FIGURE 6.5: Filtered data and spectrum.

Change the x-axis from time to the saw-tooth signal V(Sweep) to display the eye as in Fig. 6.6 (the unfiltered PRBS data eye display is also displayed on the left). Set the noise generator by adjusting the gain which should result in the distorted eye pattern as shown.

The eye diagram measures intersymbol interference (ISI) that is defined as the ratio of the eye sizes as

$$h = 3.86,\ H = 5.8,\ \text{so ISI} = -3.5\ \text{dB}. \tag{6.1}$$

The smaller eye height is h, and H is the maximum eye opening. A jittered clock is achieved using a **FILESTIM1** part that reads in a jittered clock signal. Change the PRBS clock to the jittered clock and notice how the eye becomes distorted as demonstrated in Fig. 6.7.

6.3 VECTOR/SCATTER DIAGRAM

A scatter diagram is a useful diagnostic tool for assessing the performance of modems. This diagram is a sampled version of the eye diagram, with sampled Q and I signals displayed orthogonally. The effects of noise, time offset, frequency offset, phase offset, etc. for BPSK and QPSK may be observed in a scatter diagram. The circuit in Fig. 6.8 is a schematic for producing 8-PSK and shows a scatter instrument attached to each BM output.

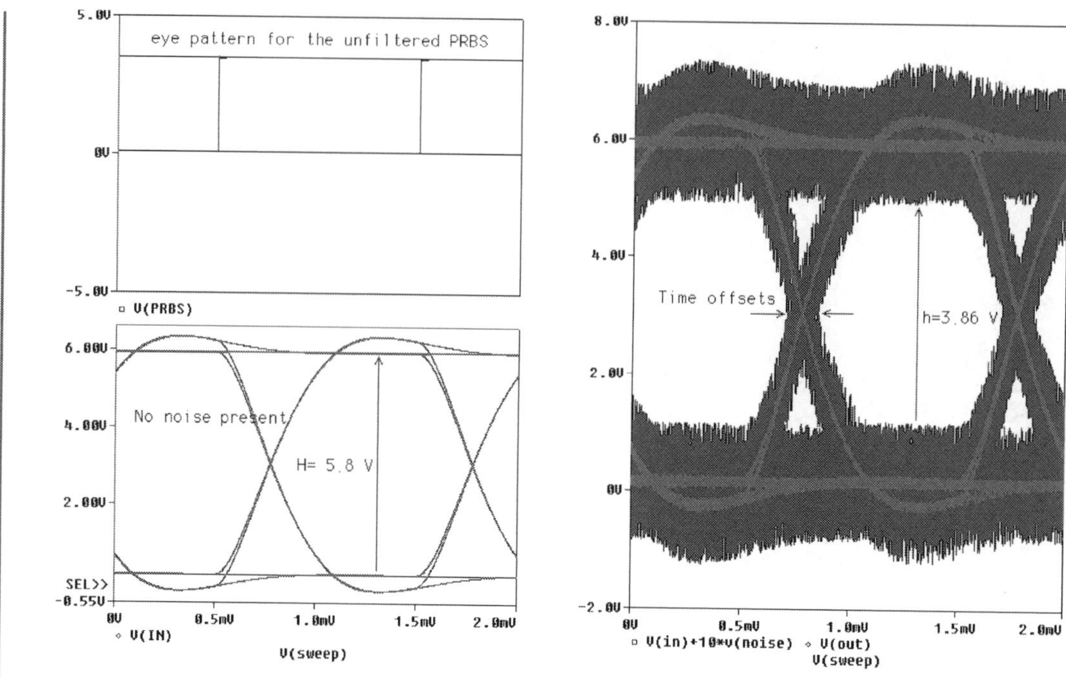

FIGURE 6.6: Eye diagram.

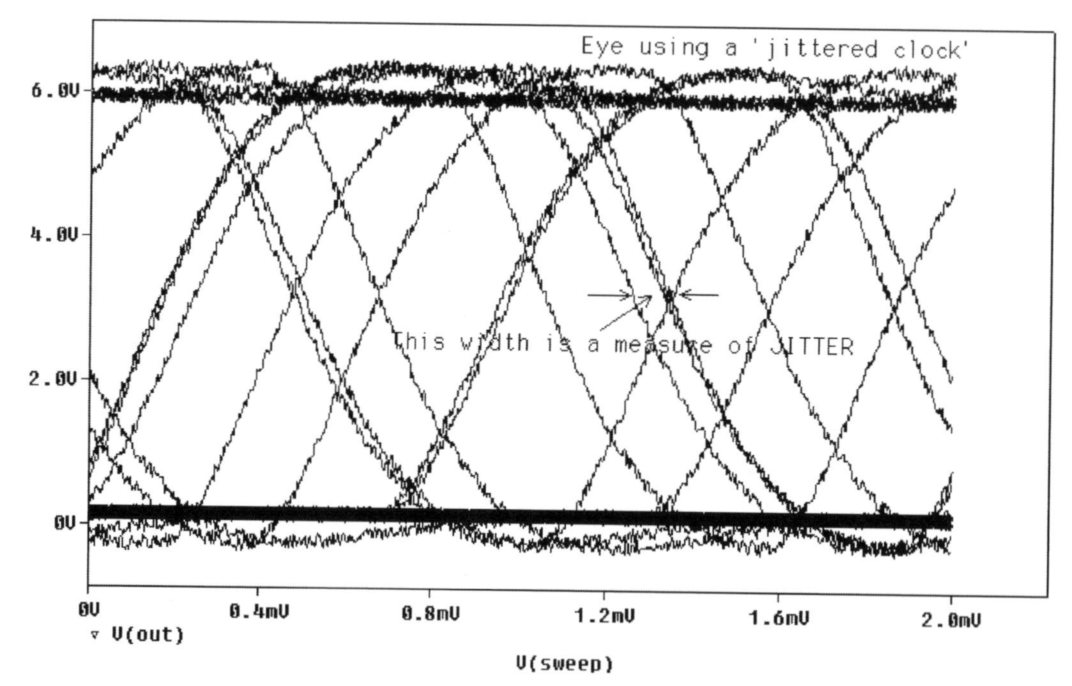

FIGURE 6.7: Distorted eye diagram due to clock jitter.

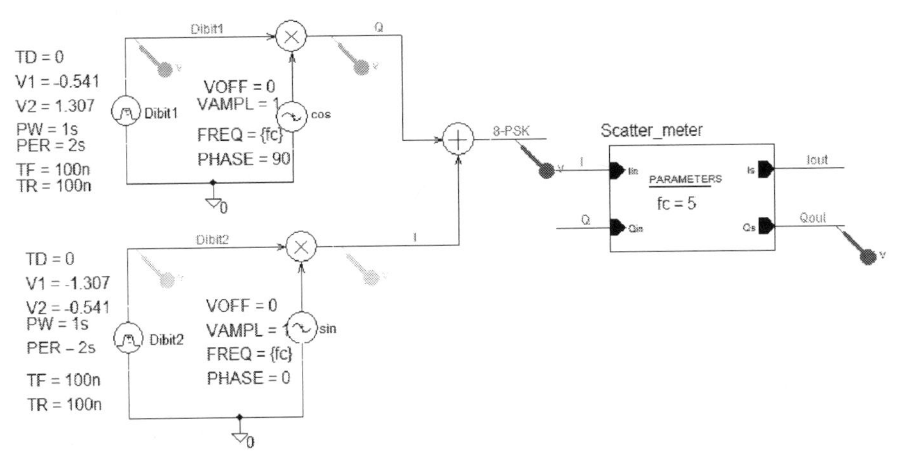

FIGURE 6.8: 8-PSK and the scatter instrument.

Fig. 6.9 shows the inside of the scatter meter. Two switches sample the *I* and *Q* output lines and the sampler parameters TD, PW and PER are a function of the carrier frequency which is defined externally in a PARAM part.

Set the **Output File Options/Print values in the output file** to 1 s, **Run to time** to 10 s, and **Maximum step size** to 1 μs. Simulate with **F11** key. The tribit input voltage levels shown in Fig. 6.10 produce corresponding carrier phase changes.

The scatter meter assesses the performance of *M*-ary systems and needs to be connected to parts of the circuit under investigation. In this first application we attach the meter to the QPSK system investigated in a previous chapter. The QPSK constellation diagram shown in Fig. 6.11 was obtained by changing the *x*-axis from time to V(Iout) and plotting the quadrature signal V(Qout) on the *y*-axis when in **PROBE** after simulation.

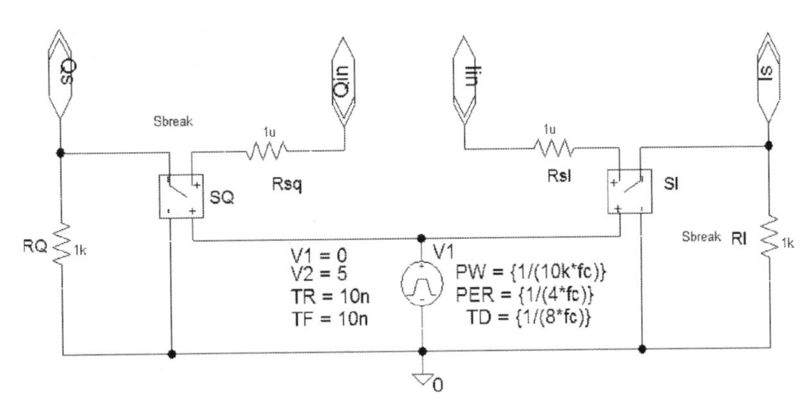

FIGURE 6.9: Inside the scatter instrument.

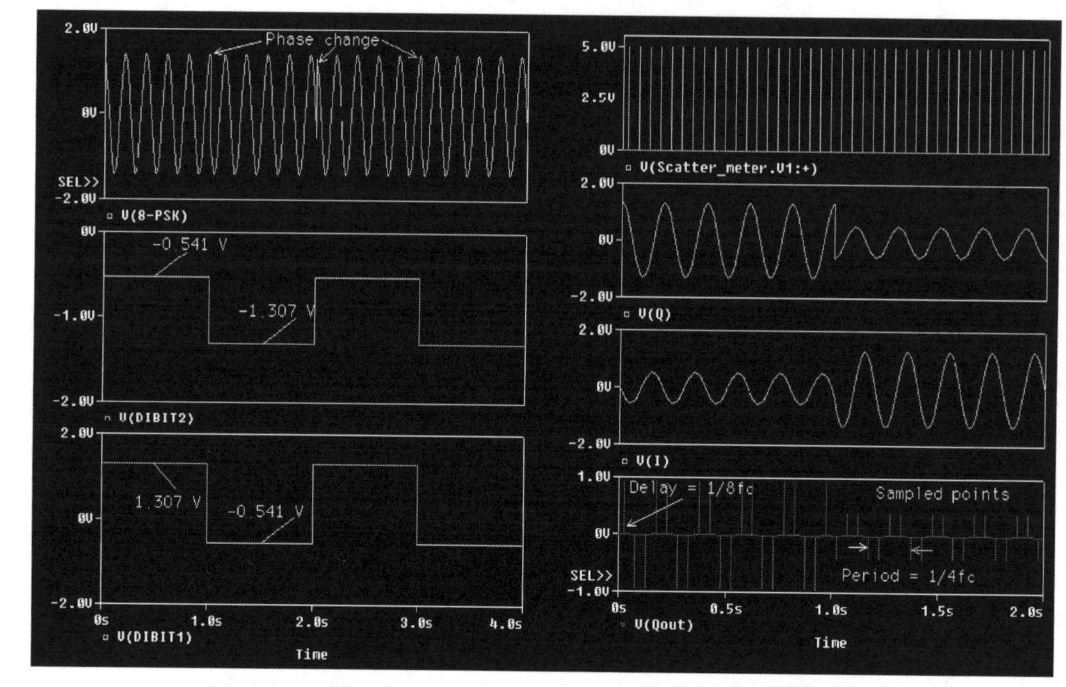

FIGURE 6.10: Test data for 8-PSK.

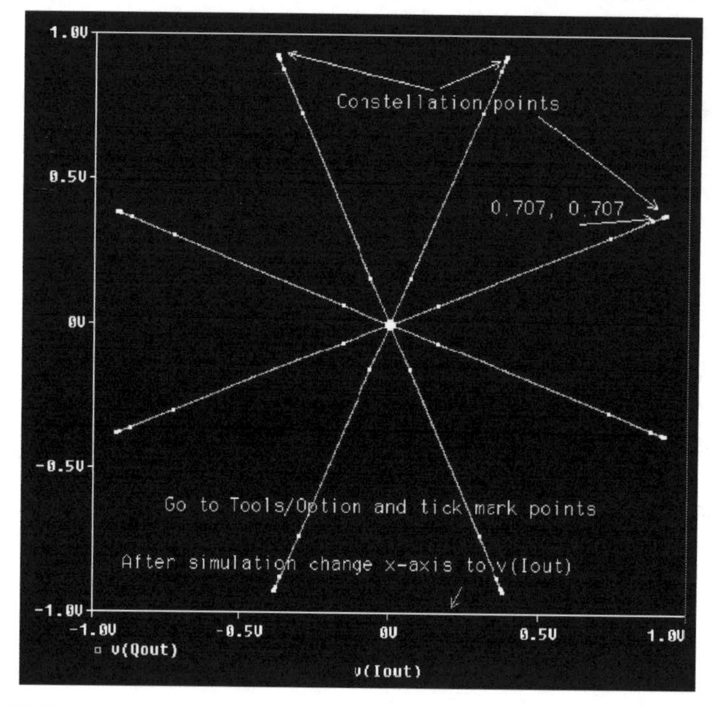

FIGURE 6.11: QPSK constellation.

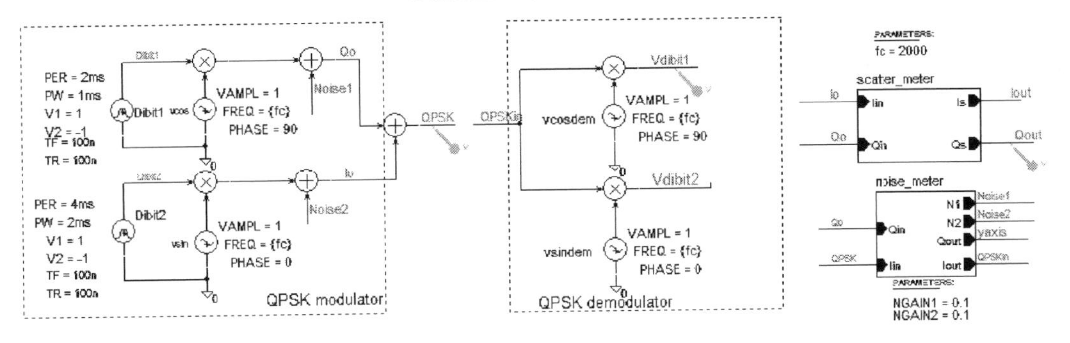

FIGURE 6.12: Adding noise to the QPSK signal.

The scatter instrument is attached to the Q and I outputs to produce the constellation diagram. A more suitable **PROBE** display is set by selecting **Tools/Options** and ticking **Mark Data Points**. The aspect ratio might need to be reset by resizing the vertical side of the plot using the mouse on the right- or left-hand side of the display.

6.3.1 Noisy QPSK Scatter Diagram

To investigate the effect of noise on the constellation, we need to use the noise meter which has two different noise source outputs noise1 and noise2. These two sources are connected to the Qo and Io outputs as shown in the QPSK circuit in Fig. 6.12.

Fig. 6.13 shows how the vector points are deviated from their correct location to another point. Thus, in a high-density M-ary PSK system, a noisy point could be interpreted incorrectly by the demodulator. Vary the noise amplitude and investigate the effect on the constellation using a scatter meter.

6.3.2 Noisy 16-QAM Scatter Diagram

Another application is to attach the scatter and noise meters to the output of the 16-QAM schematic discussed in chapter 5 (see Fig. 5.28) to display the noisy M-ary signal scatter diagram as shown in Fig. 6.14.

6.4 CLOCK WITH JITTER

Noise, limited bandwidth, and jitter degrade the quality of signals sent over a communication channel. We have already considered the effects of noise and bandwidth on the signal, so what remains is to investigate how jitter affects the horizontal components of the eye diagram. Jitter occurs in two forms: random jitter due to noise and deterministic jitter due to signal modulation by mains interference etc. Fig. 6.16 is a schematic to create a jitter text file using an external noise source to modulate, in a random fashion, the leading edge of the clock produced

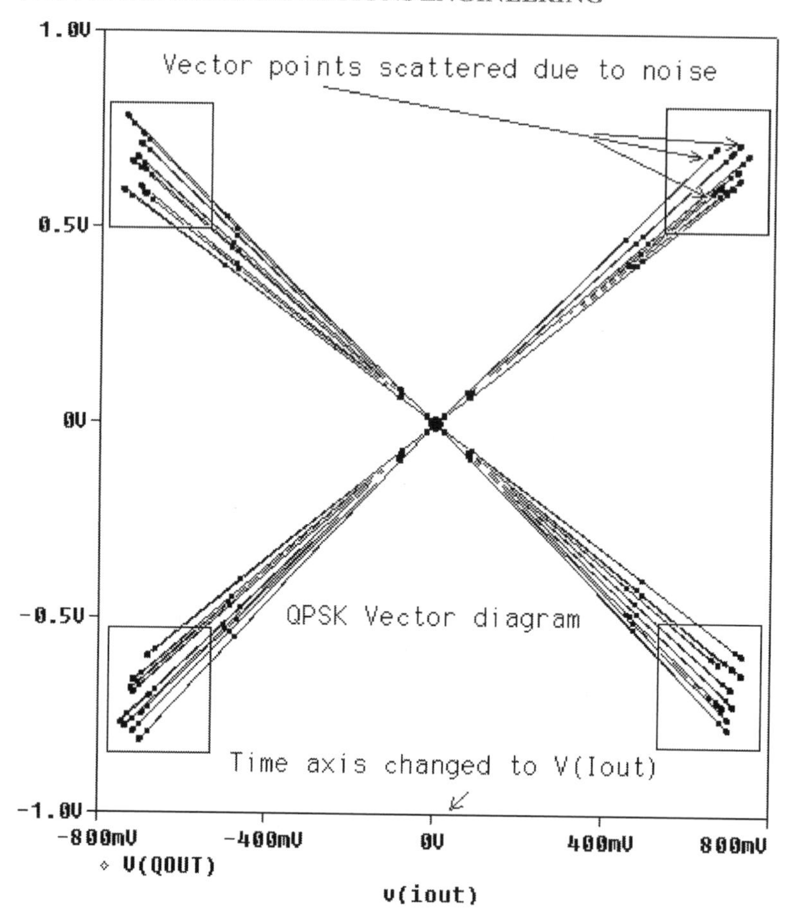

FIGURE 6.13: Noisy constellation diagram.

by the 555 IC. The clock frequency of the VCO for a 555 IC clock frequency is determined $f_0 = \frac{1.44}{(R_1+2R_2)C_1}$. A 50% duty cycle is achieved making $R_2 \gg R_1$ with the mark–space ratio set as $[(R_1 + R_2)/(R_1 + 2R_2)] \times 100$.

The jitter text file was created by attaching a VECTOR part to the circuit as shown. This records the digital signal at this point. High-light the vector part and select **Edit Properties** and enter the name and location where the file is to be recorded. The noise text file "info.txt" inside the eye-meter was generated in Matlab and stored in the **signalsources** directory where it is applied using a **VPWL_F_RE_FOREVER** generator part. The amount of jitter depends on the noise amplitude set by the **noiseamplitude** parameter defined in the **PARAM** part. In Fig. 6.4 the standard **DIGCLOCK** part was replaced by the new "jittered" clock to observe the effect of jitter on the eye diagram. After simulation, change the x-axis variable to V(sweep) as shown.

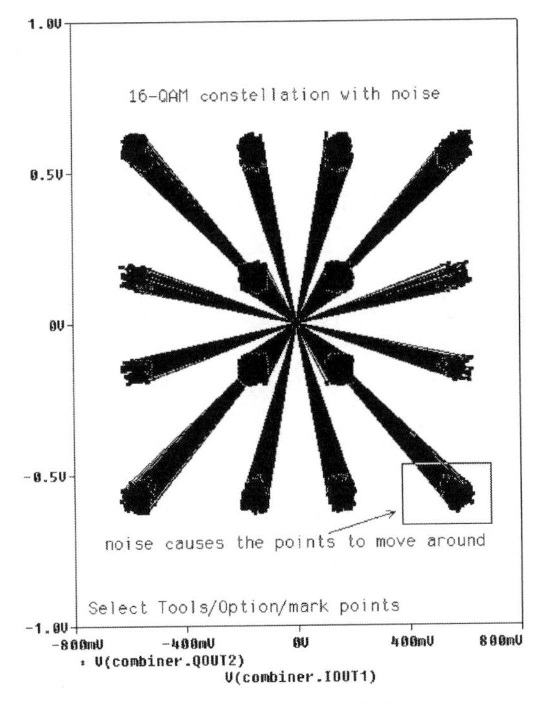

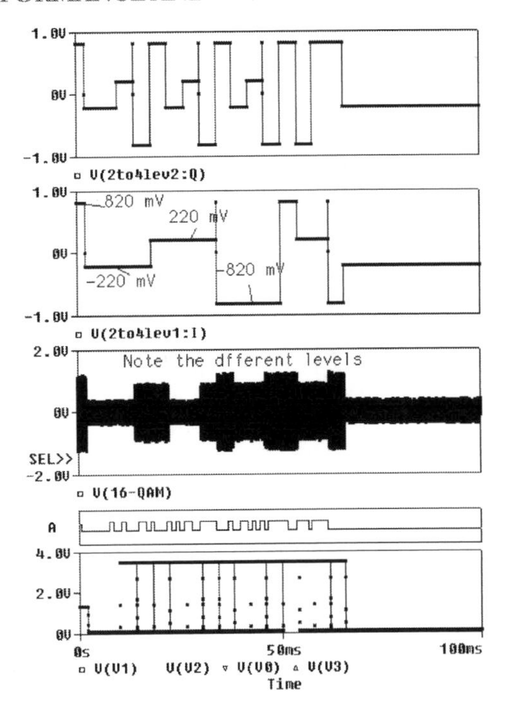

FIGURE 6.14: 16-QAM constellation.

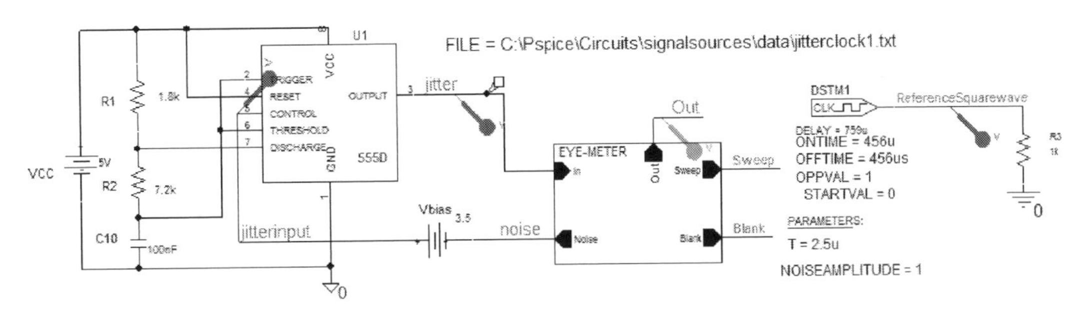

FIGURE 6.15: Jitter generation.

6.5 INTERSYMBOL INTERFERENCE

Fig. 6.17 is a PRBS generator for creating random transmitted data and is applied to a transmission line.

The transmission line behaves like a low-pass filter and distorts the sequence causing ISI, as shown in Fig. 6.18. Transmission lines have a delay and as such cause interference between the delayed pulses on the line.

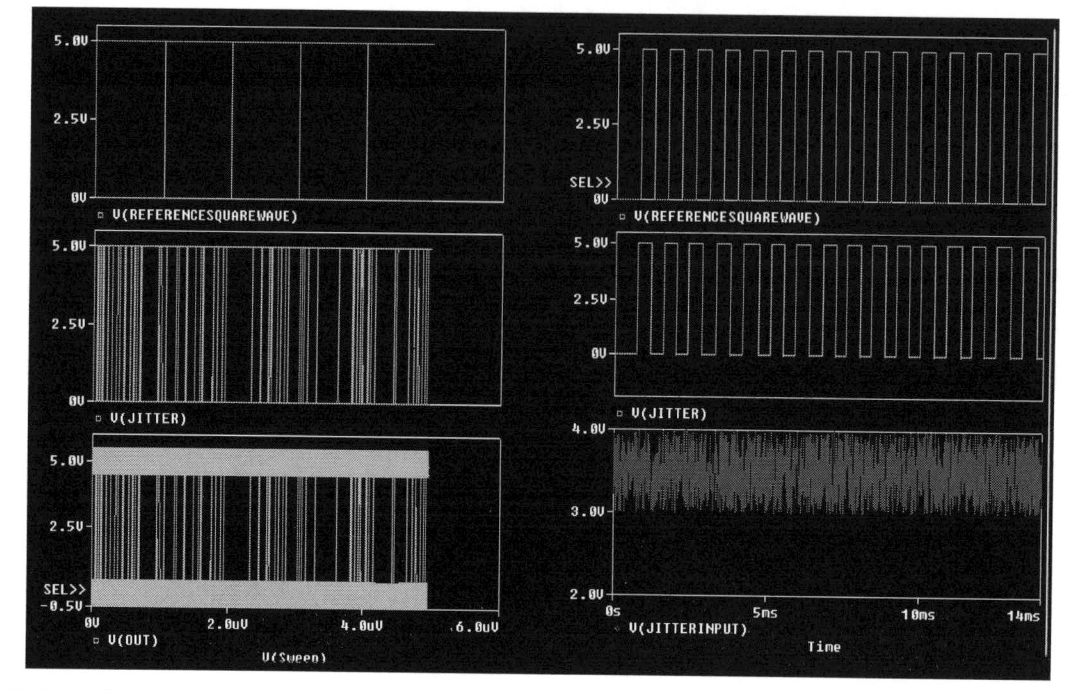

FIGURE 6.16: Eye diagram showing clock jitter.

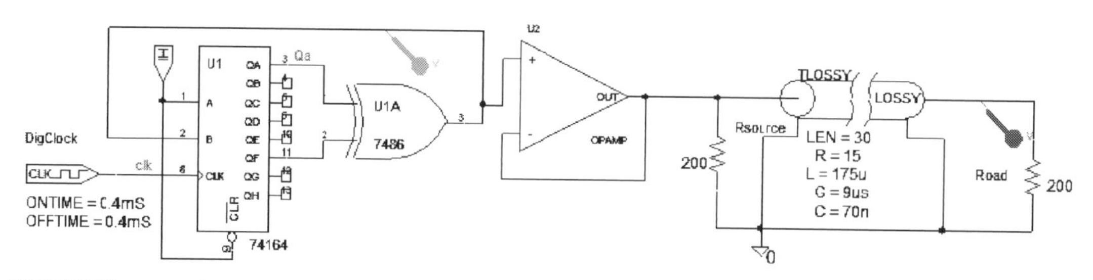

FIGURE 6.17: ISI production.

6.6 NYQUIST SIGNAL CRITERION

Rounding the edges of a rectangular pulse, of width τ, causes most of the pulse frequency spectrum to be concentrated within the first main lobe (i.e., 0 and $1/\tau$). Frequencies outside this region are much smaller compared to rectangular pulses. Thus most of the transmitted signal power is within a bandwidth of $1/\tau$. In the receiver, correct decisions about individual pulses are made provided pulses are not overlapping. Applying the inverse Fourier transform to such a pulse produces an impulse response defined as

$$h(t) = \frac{1}{T}\frac{\sin(\pi t/T)}{(\pi t/T)}. \tag{6.2}$$

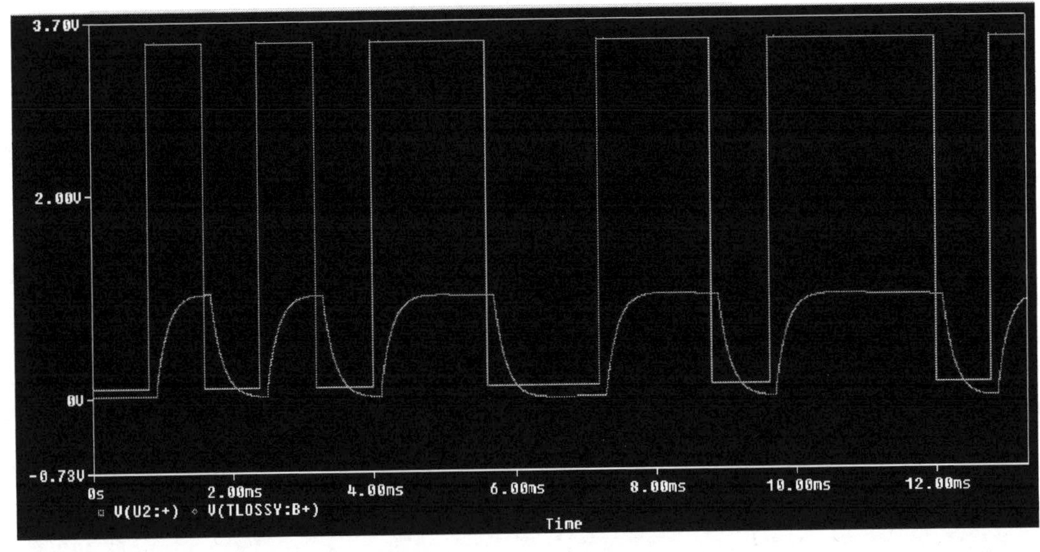

FIGURE 6.18: The PRBS sequence distorted by the transmission line.

However, generating ideal sinc signals with unlimited tails extending to infinity is unrealizable and non-causal (attempts to exist before time equal to zero). However, we can fix this by introducing a delay factor. To display the left hand tails in a non-causal manner, we must introduce a time shift $k^{*}T$ as shown in (6.3), where k is an integer and T is the sampling pulse width.

$$h(t) = \frac{1}{T}\frac{\sin(\pi(time - k * T)/T)}{(\pi(time - k * T)/T)}. \tag{6.3}$$

However, let us first investigate sinc production with the factor kT set to zero initially as shown in Figure 6.19. Here, an **ABM** part defines the sinc signal and a correctly-terminated transmission line produces various delays with T set to 0.5s.

Fig. 6.20 shows superimposed waveforms with no interference at multiples of the delay time ($k = 0$).

However, let us first investigate sinc production with the factor kT set to zero initially, as shown in Figure 6.19. An **ABM** part and a correctly-terminated transmission line is employed for producing various delays with T set to 0.5s.

We will now introduce the delay factor kT to display the left hand side of the sinc signal. Change the value of T to 125 us and set $k = 8$. Set the Analysis tab to Analysis type: **Time Domain** (Transient), **Run to time** = 3ms. Press F11 to display the sinc signal shown in Fig. 6.21 that is delayed by $kT = 6^{*}125$us seconds. This produces an artificially created zero point

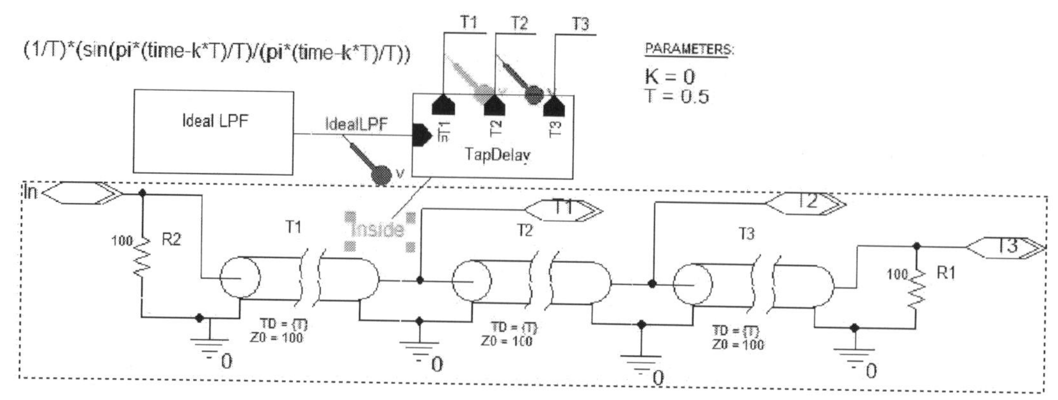

FIGURE 6.19: Ideal pulse.

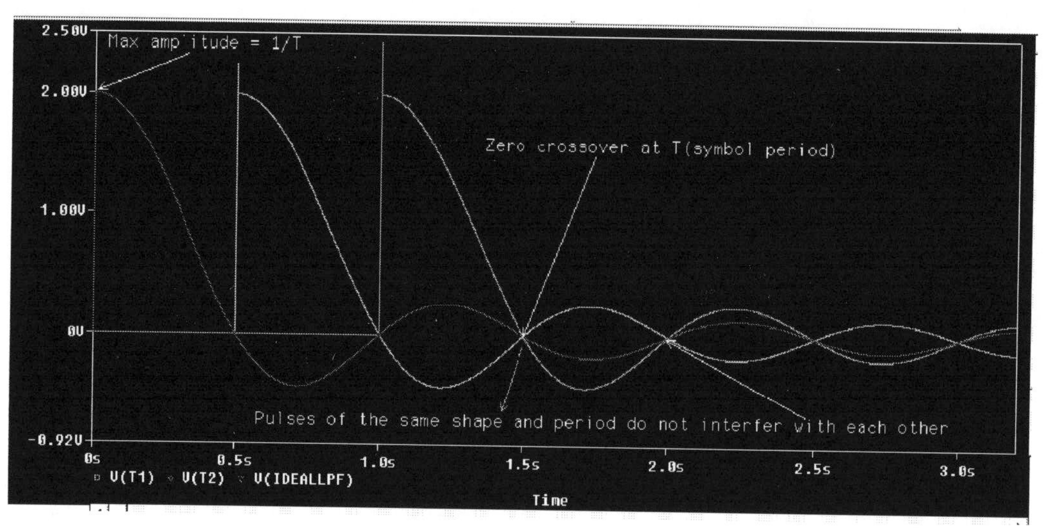

FIGURE 6.20: Sinc pulse and delayed sinc pulses.

with the signal, a type of recorded signal, displaced six symbol times to the right. Create a new plot with **alt PP** and copy the variables into the new area.

Press the **FFT** icon and observe, using the magnifying tool, the ideal low-frequency filter frequency response in the left plot. After simulation when you add notes, arrows, etc, to a **PROBE** display and you want to a permanent record of it, then from the **WINDOW** menu select **Display Control**. Type in "idealsinc" in the **New Name** box and press **Save**. To retrieve this at a later time, open **PROBE**, and from the **File/Open** menu, open Fig. 6.19.dat. Selecting this file produces a blank **PROBE** screen. Press the **WINDOWS** menu and select

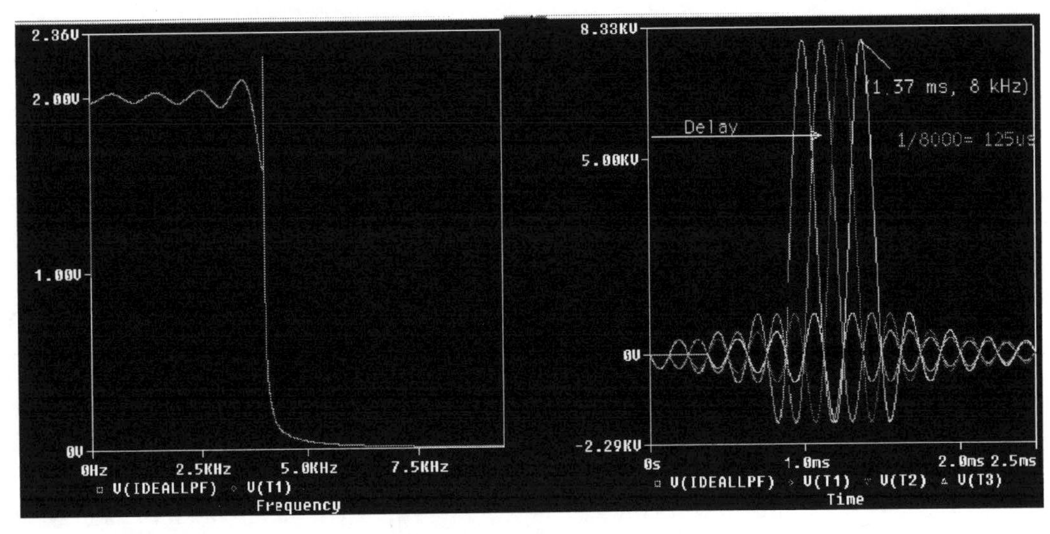

FIGURE 6.21: Sinc pulse with a 6 s advance.

Display Control menu where you should see your saved file. Select this file and press **Restore** which should restore the previous Probe plot with all text information.

6.7 RAISED COSINE FILTER

A raised cosine filter frequency response, with roll-off parameter α, is defined as

$$H(F)=\begin{bmatrix} T & \text{for} & |f| \le (1-\alpha)/(2T) \\ \frac{T}{2}\left[1+\cos\left(\frac{\pi T}{\alpha}\left\{|f|-\frac{(1-\alpha)}{(2T)}\right\}\right)\right] & \text{for} & (1-\alpha)/(2T) \le |f| \le (1+\alpha)/(2T) \\ 0 & \text{for} & (1+\alpha)/(2T) \le |f| \end{bmatrix}. \tag{6.4}$$

Here T is the symbol period and the impulse response equation is obtained using the inverse Fourier transform on (6.4):

$$h(t) = \frac{\sin c\,(\pi t/T)\cos(\pi \alpha t/T)}{T - 4\alpha^2 t^2/T}. \tag{6.5}$$

A nonrealizable brick-wall filter is produced for $\alpha = 0$ but a more realistic value is 0.4.

6.7.1 Square Root-Raised Cosine Filter

The square root-raised filter (RCC) located at the transmitter output and receiver input is so-called because we take the square root of the equation that defines the frequency response given by:

$$H(f) = \begin{bmatrix} 1 & \text{for} & |f| < f_N(1-\alpha) \\ \left\{0.5 + 0.5 \sin \dfrac{\pi}{2f_N}\left(\dfrac{f_N - |f|}{\alpha}\right)\right\}^{1/2} & \text{for} & f_N(1-\alpha) \le |f| \le f_N(1+\alpha) \\ 0 & \text{for} & |f| > f_N(1+\alpha) \end{bmatrix}. \tag{6.6}$$

The square root-raised cosine impulse response, for a symbol period T, is

$$g(t) = \frac{\frac{4\alpha t}{T}\cos[\frac{\pi t}{T}(1+\alpha)] + \sin[\frac{\pi t}{T}(1-\alpha)]}{\frac{\pi t}{T}[1 - (\frac{4\alpha t}{T})^2]}. \tag{6.7}$$

This translates into the following equation when we enter it into an **ABM** part (see Fig. 6.23) as ((4*a*(time-k*T)/T)*cos((1+a)*pi*(time-k*T)/T)+sin((1-a)*pi*(time-k*T)/T))/((pi*(time-k*T)/T)*(1-PWR((4*a*(time-k*T)/T),2))). The Nyquist frequency is $f_N = \frac{1}{2T_s} = \frac{R_s}{2}$ where R_s is the symbol rate. The frequency response for the transmitter and receiver root-raised cosine filters, when taken as a combined unit, produces a raised cosine filter response.

6.7.2 Raised Cosine Filter Response

Chapter 1 showed how synthesized and square wave signals are affected when they are low-pass filtered by the transmission line. Pulses sent down such a channel are attenuated and phase distorted due to a limited bandwidth. These two factors, together with timing inaccuracies, produce intersymbol interference (ISI). Thus, correct decisions about the received pulse amplitude cannot be made in the receiver when pulses are overlapping and it causes errors. To reduce ISI, a raised cosine filter is located at a transmitter output stage and rounds off the edges of data pulses prior to transmission, thus concentrating most of the pulse energy within the main spectral lobe (i.e., between 0 and $1/\tau$). A similar filter is located at the receiver input. The filter gain is constant in the passband region and gets its name because of the cosine shape in the roll-off region and is symmetrical about f_c as shown in Fig. 6.22.

The excess bandwidth r over f_c reduces the ripple duration in the impulse response. The theoretical maximum transmission rate, or Nyquist rate, is $2W$ pulses per second for a channel of bandwidth W Hz, if pulses are to be recovered in the receiver. A useful roll-off range is 0.2–0.4 and is related to the minimum available bandwidth as

$$W = W_{\min}(1+\alpha). \tag{6.8}$$

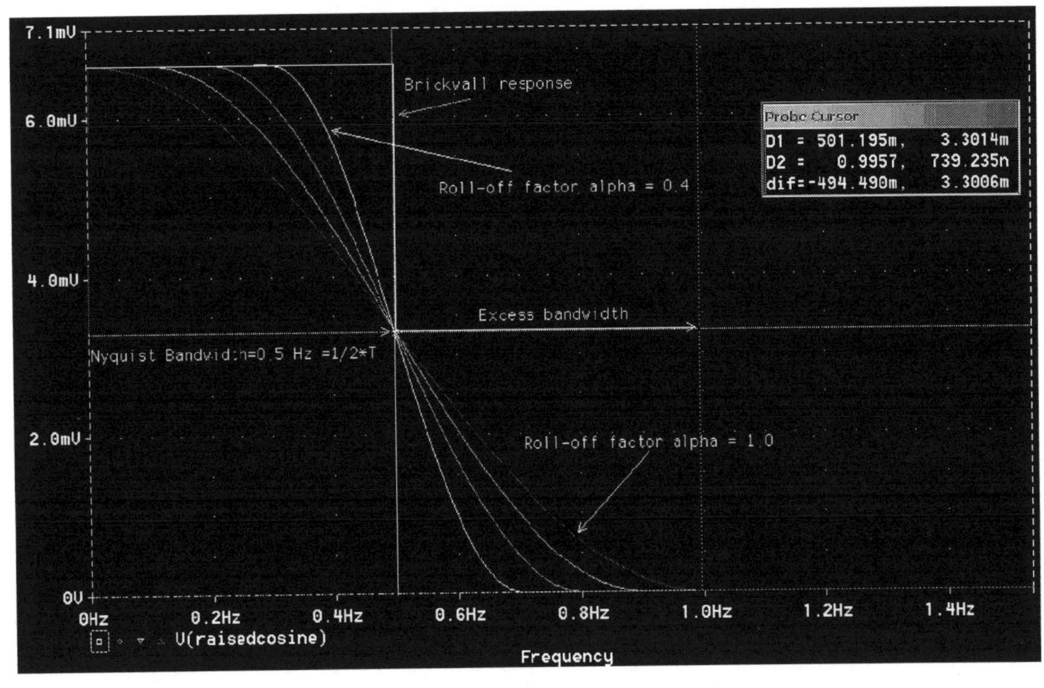

FIGURE 6.22: Raised cosine filter frequency response.

We may express (6.8) for maximum ISI avoidance in terms of the symbol transmission rate R_B (for rectangular-shaped pulses), and bandwidth W:

$$W = \frac{1}{2T}(1+\alpha) = \frac{R_B}{2}(1+\alpha) = f_c(1+\alpha). \tag{6.9}$$

The Nyquist rate has a roll-off factor $\alpha = 0$, so $R_B = 2\,\mathrm{W}$. Manipulate (6.9) to yield

$$W = (1+\alpha)fc. \tag{6.10}$$

6.7.3 Transmitter and Receiver Filter Impulse Response

Fig. 6.23 investigates the family of raised cosine filters, with each filter impulse response equation defined in an **ABM** part. The raised cosine function has $1/T^2$ in the denominator and is responsible for the response tails falling off more rapidly when compared to the sinc function and results in smaller ISI. To investigate the impulse and frequency responses, we need a **PARAM** part to define the roll-off parameter, α, that is varied from 0.4 to 1. To vary α, select **Analysis Setup/Parametric** and set **Name** = a, **Start Value** = 0.4, **End Value** = 1, and **Increment** = 0.2. The transient parameters: **Output File Options/Print values in the output**

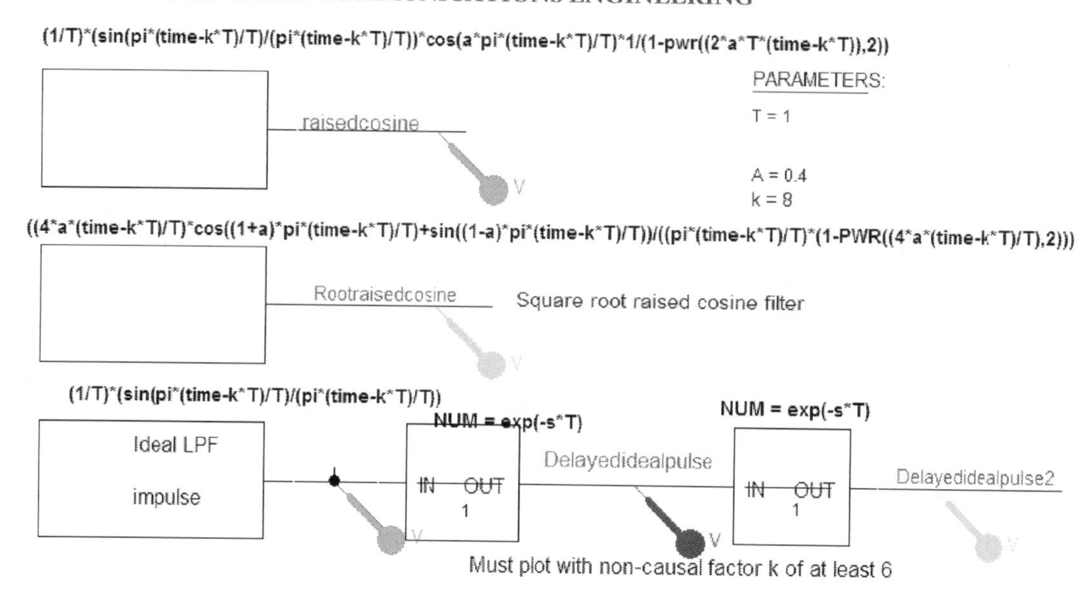

FIGURE 6.23: Raised cosine filter impulse response production.

file = 1 s, **Run to time** = 300 s, and **Maximum step size** = 10ms. Place markers on each filter output and simulate.

When the response appears, use the magnifying glass icon to select the first 20 s from the 300 s display. We need a large **Run to time** for good **FFT** resolution. The raised cosine signal for different roll-off factors is displayed in Fig. 6.24. For the ideal case $\alpha = 0$, so $R = 2$ W. Press the **FFT** icon in **PROBE** to observe the raised cosine frequency response that is flat in the passband region but has a sinusoidal shape in the transition region. The response is symmetrical about the cut-off frequency f_c (voltage gain of 0.5, or -6 dB at f_c) and requires an excess bandwidth of α over f_c which greatly reduces the ripple duration in the impulse response. The theoretical maximum Nyquist rate for transmitting pulses over a channel with bandwidth W and recovering them exactly is $2W$ pulses per second. The bandwidth extends from W to $2W$. The roll-off factor shows how much more bandwidth is needed when compared to the ideal Nyquist bandwidth. This is 100% for $\alpha = 1$ (a large value means large bandwidth). Fig. 6.24 shows impulse and frequency responses for the cosine family of filters.

To display the frequency response, select **PROBE/Window/New** and **Tile Vertically**. In the new window, apply **alt PP** three times to give three separate plots and then copy the three variables across by selecting the variables at the bottom. Press the **FFT** icon but make sure that the blue bar at the top of the window plot is activated.

The -3 dB cut-off frequency is measured where the output is equal to the passband gain/sqrt(2) = 6.66*0.707 = 4.7 V. Compare this unsymmetrical response to the raised-cosine

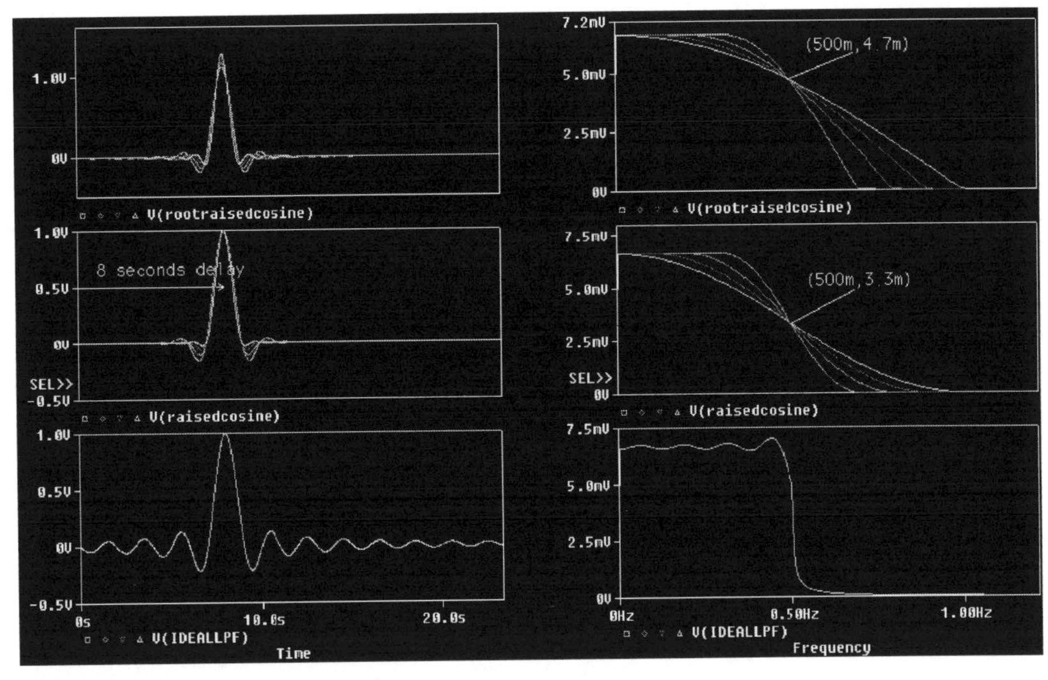

FIGURE 6.24: Cut-off frequency equals 0.5 Hz.

response. The cut-off frequency, f_c, for the root-raised response is measured at −3 dB but is −6 dB for the raised-cosine filter, the difference being due to the square root factor.

Set the transient analysis parameters and simulate. To plot the root-raised and raised cosine transient responses in decibels as shown in Fig. 6.25, use the **dB()** function from

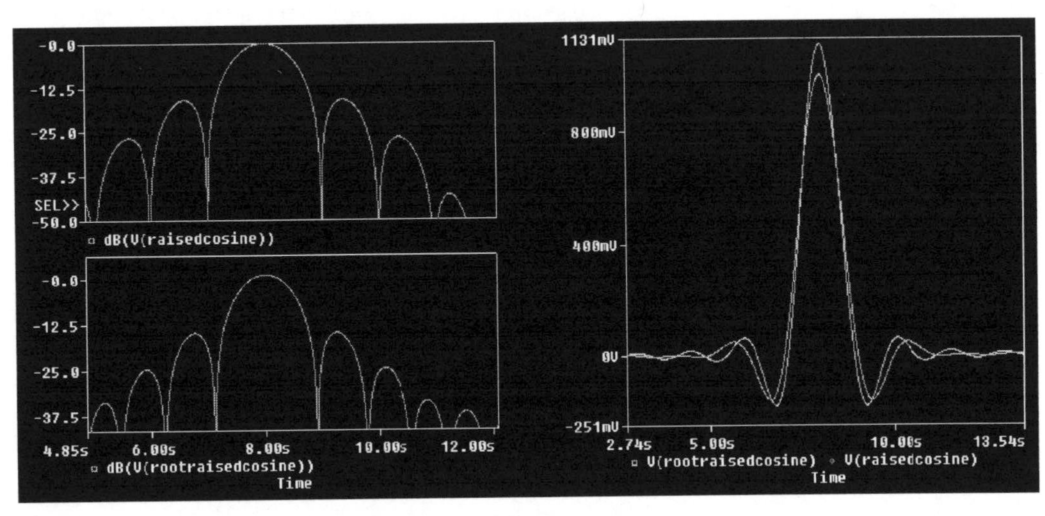

FIGURE 6.25: Raised and root cosine filter responses.

the list of functions available from the **Trace** menu in Probe. Select the plotted variable **V(raisedcosine)** and in the **Trace Expression** box enter **dB(V(raisedcosine))**. Repeat for the **V(rootraisedcosine)** variable.

6.7.4 Example

(i) A mobile phone contains a raised cosine filter to reduce intersymbol interference (ISI). The baseband bandwidth is 15 kHz, a 30 kHz passband bandwidth and $\alpha = 0.25$. Solution: The maximum symbol rate R_B to achieve zero ISI is $R_B = 2W/(1+\alpha) = 2 \times 15 \times 10^3/(1+0.25) = 24$ kbaud giving a bit rate for eight M-ary encoding as 8×24 baud $= 192$ kbits/s. (ii) A telephone passband system, with a 2400 Hz (600–3000 Hz) channel bandwidth, is equalized using a raised cosine filter. A 2400-bit/s quadrature phase shift keying (QPSK) transmission system has a mid-channel frequency $f_c = 1800$ Hz. The signal spectrum is accommodated when the roll-off factor α is unity. Calculate the absolute and 6 dB bandwidths and the roll-off factor to fit a 4800-bit/s 8-PSK signal into the available channel spectrum. Will the signal spectrum be accommodated for unity roll-off factor? Find the absolute and 6-dB bandwidths.

6.7.5 Solution

QPSK has four phase changes, so $M = 4 = 2^n \Rightarrow n = 2$, and the baud rate is $2400/2 = 1200$ baud. The absolute bandwidth from (6.9) is

$$R_B = \frac{2W}{1+\alpha} \Rightarrow W = \frac{R_B(1+\alpha)}{2}. \qquad (6.11)$$

For $\alpha = 1$, and $R_B = 1200$, $W = 0.5(1200(1+1)) = 1200$ Hz. The absolute bandwidth $B = 2W = 2400$ Hz thus fits into the available bandwidth. The relationship between B, the absolute bandwidth, and the -6 dB cut-off frequency f_c for raised cosine filtering is

$$f_\Delta = B - f_c. \qquad (6.12)$$

The roll-off rate and cut-off frequency are related:

$$f_\Delta = \alpha f_c \Rightarrow \alpha = f_\Delta / f_c. \qquad (6.13)$$

If $\alpha = 1$, then $f_\Delta = f_c$, so that $f_c = B - f_c \Rightarrow B = 2f_c = 2400$ Hz. 8-PSK has eight levels ($M = 8 = 2^n$ levels), so for 4800-bit/s, $n = 3$, the baud rate is $R = 4800/3 = 1600$ baud, and $W = 1600(1+1)/2 = 1600$ Hz $\Rightarrow B = 2W = 3200$ Hz, which exceeds the 2400 Hz available bandwidth. A raised cosine filter ensures that the signal does not exceed the available channel bandwidth and has a bandwidth $B = 2W = 2400$ Hz, so $W = 1200$ Hz.

The roll-off factor is

$$R_B = \frac{2W}{1+\alpha} \Rightarrow r = \frac{2W}{R_B} - 1 = \frac{2400}{1600} - 1 = 0.5. \quad (6.14)$$

The relationship between the absolute bandwidth, B, and the -6 dB cut-off frequency is

$$f_\Delta = B - f_c \Rightarrow B = f_\Delta + f_c. \quad (6.15)$$

The relationship between the roll-off factor, α, and the -6 dB cut-off frequency, is

$$B = f_\Delta + f_c = f_c(f_\Delta / f_c + 1) = f_c(\alpha + 1) \quad (6.16)$$

$$f_c = \frac{B}{1+\alpha} = \frac{1200}{1.5} = 800\ Hz. \quad (6.17)$$

A channel bandwidth W has a maximum signaling rate of $2W$ pulses per second for no errors, the right pulse shape, and an ideal flat channel frequency response (i.e., no distortion and no noise). In M-ary signaling, each symbol represents one of M different levels to convey $\log_2 M$ bits of information, and produces a maximum bit rate of $2W \log_2 M$ bps. The bandwidth efficiency is a ratio of the bit rate R to the bandwidth W:

$$\frac{R}{W} = 2 \log_2 M. \quad (6.18)$$

6.8 ERRORS, NOISE, AND MATCHED FILTERS

The function of a matched filter in a receiver is to minimize the probability of errors. Thermal noise power is

$$P_{\text{noise}} = kTB = N, \quad (6.19)$$

where k = Boltzmann's constant $= 1.38 \times 10^{-23}$ J/K, T = resistor temperature in kelvin (K) $=$ 273 + °C and B = bandwidth in Hz. Band-limiting a receiver signal limits the noise to a value consistent with good signal quality. Noise is called white since it has a spectrum extending from 0 Hz to 10^{13} Hz (a range similar to white light) and has a constant power spectral density measured per hertz of bandwidth (kT W/Hz). N_0 is the amount of noise in each Hz of the transmission band, and is expressed in units of W/Hz having a range between 10^{-7} and 10^{-21} W/Hz. From (6.19), $N = kTB$, thus $N_0 = N/B$, or $N = N_0 B$. To quantify the signal present to the amount of noise present, we must define a ratio of signal power to noise power (S/N ratio or SNR) in decibels, as SNR $= 10 \log_{10} S/N$.

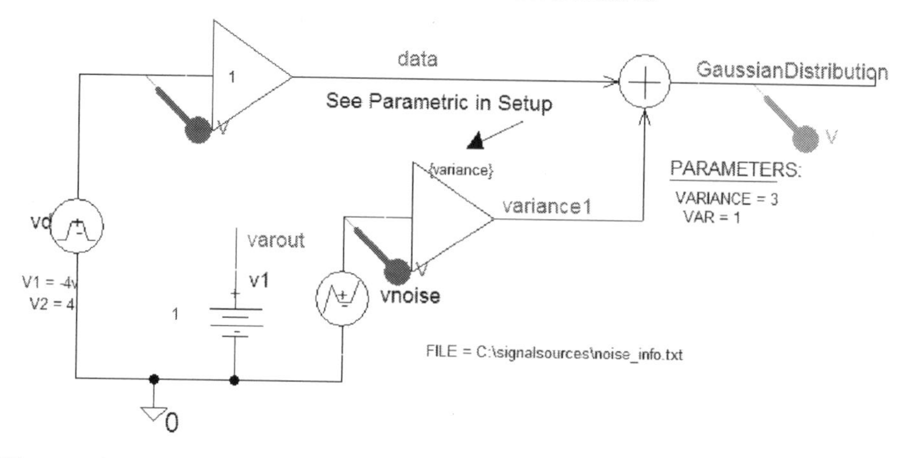

FIGURE 6.26: Gaussian distribution for two events with variable noise variance.

6.8.1 Importing Noise into a Schematic

The schematic shown in Fig. 6.26 imports an ASCII noise file comprising time–voltage pairs created in Matlab. The output wire segment is called **GaussDistribution**. Different amounts of noise can be added to the NRZ-B data signal to investigate the change in the signal to noise ratio.

The +4 V signal represents logic 1 and −4 V is logic 0. The threshold, or decision level, for deciding whether 1 or 0 is present is set to 0 V (i.e., the average of the two signals). Set the **Transient Analysis** parameters as follows: **Output File Options/Print values in the output file** = 1 µs, **Run to time** = 5ms, **Maximum step size** = (blank), and press **F11** to simulate. Fig. 6.27 shows the occasional noise positive signal to dip below the zero threshold decision, and the negative signal to rise above the zero threshold level. At these instances, the receiver misinterprets the input signal and thus over this 5ms period produces errors. To quantify these errors we need to look at the probability of these errors existing over a certain time period.

6.8.2 Gaussian Noise Distribution Plot Using a Macro

A macro contains a sequence of instructions or equations to be run in **PROBE** to carry out a task. This can be made global and available for use in other schematics. We will illustrate how to create a macro by plotting a Gaussian distribution of noise.

Noise in telecommunication systems is assumed to have a Gaussian probability density distribution described as

$$P_e(v_d, v_n) = \frac{1}{v_n\sqrt{2\pi}} e^{-(v_n - v_d)^2/2v_n}. \qquad (6.20)$$

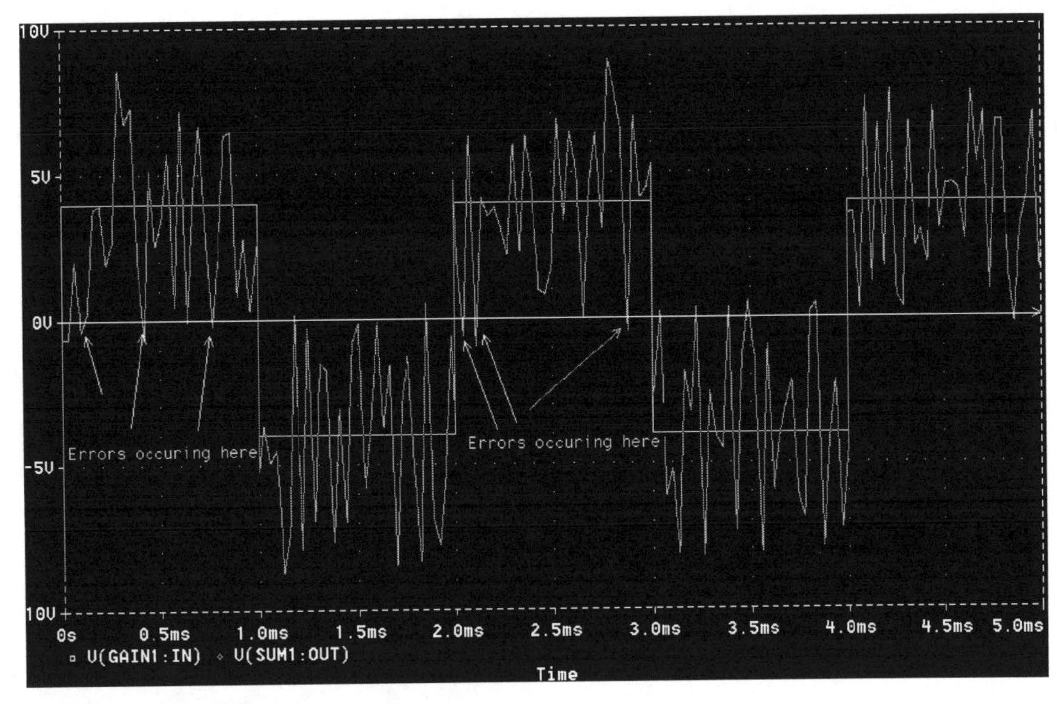

FIGURE 6.27: Data signal corrupted by noise.

The mean value of the data signal without noise is $v_d = \pm 2$ V and v_n is the RMS noise voltage. To plot the Gaussian distribution centered on the mean of the data voltages (i.e. −2 V and +2 V), we need to create two macros that are run from **PROBE\Trace\Macros** menu. In the **Definition** box, paste the first of the two equations given below, **Save** and **Close**. Repeat for the second equation. You can also open up a text editor such as Notepad and enter the Expressions:

Gaussian1(vd,Vn) =
(1/(sqrt(rms(Vn)*rms(Vn)*2*pi)))*exp(-((vd-2)*(vd-2)/(2*rms(Vn)*rms(Vn))))
Gaussian2(vd,Vn) =
(1/(sqrt(rms(Vn)*rms(Vn)*2*pi)))*exp(-((vd+2)*(vd+2)/(2*rms(Vn)*rms(Vn))))

The only difference in these two equations is the noise voltage (vd-2) and (vd+2) locating the plots around −2 V and +2 V. To plot the two distributions we need carry out the following procedure. Remove all markers from the schematic. In the **Analysis Setup** menu, tick **DC Sweep** and **Swept Var Type** as a Voltage source. In the **Name** box type in v1, **Start Value** = −5 V, **End Value** = 5 V, and **Increment** = 0.01. What we are doing here is to sweep the DC source "v1," from −5 V to +5 in steps of 0.01 V, thus simulating the input variable range.

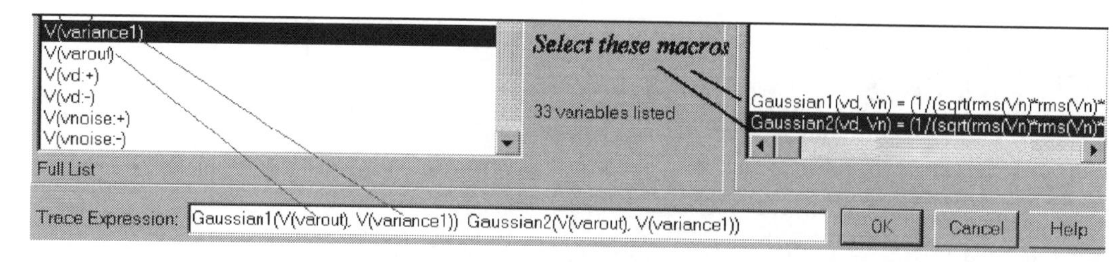

FIGURE 6.28: Entering the two Gaussian macro variables.

Press **F11** to simulate. Select the **Trace Add** icon, and at the top right-hand side in the **Functions or Macros** drop-down menu, select **Macros**. Highlight the desired macro from the list and it will be placed in the **Trace Expression** box as shown in Fig. 6.28. Replace the general data variable, *vd*, with the variable varout, and the general noise variable, *vn*, with the variable called variance1, both selected from the left list as shown. This results in the equation appearing as Gaussian1(V(varout), V(variance1)). Repeat for the second macro Gaussian2(V(varout), V(variance1)) but leave a space between the two macros.

This produces the two Gaussian distributions shown in Fig. 6.29. When the two plots intersect it generates a gray area where errors occur. Thus, the tail end of the right plot (a "1") intersects the left plot and could be interpreted as a "0".

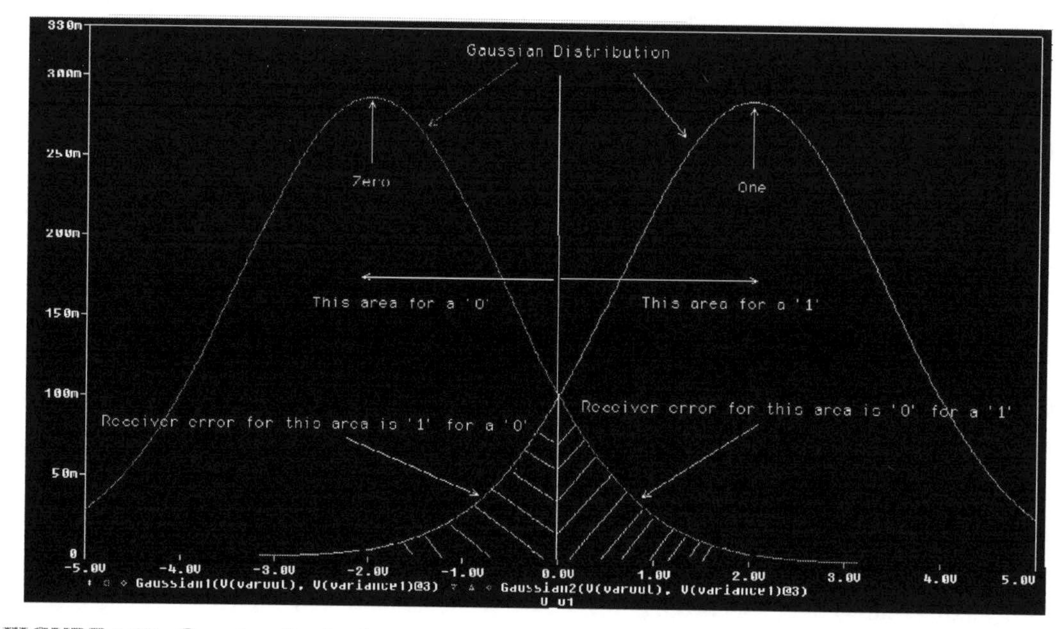

FIGURE 6.29: Gaussian distribution.

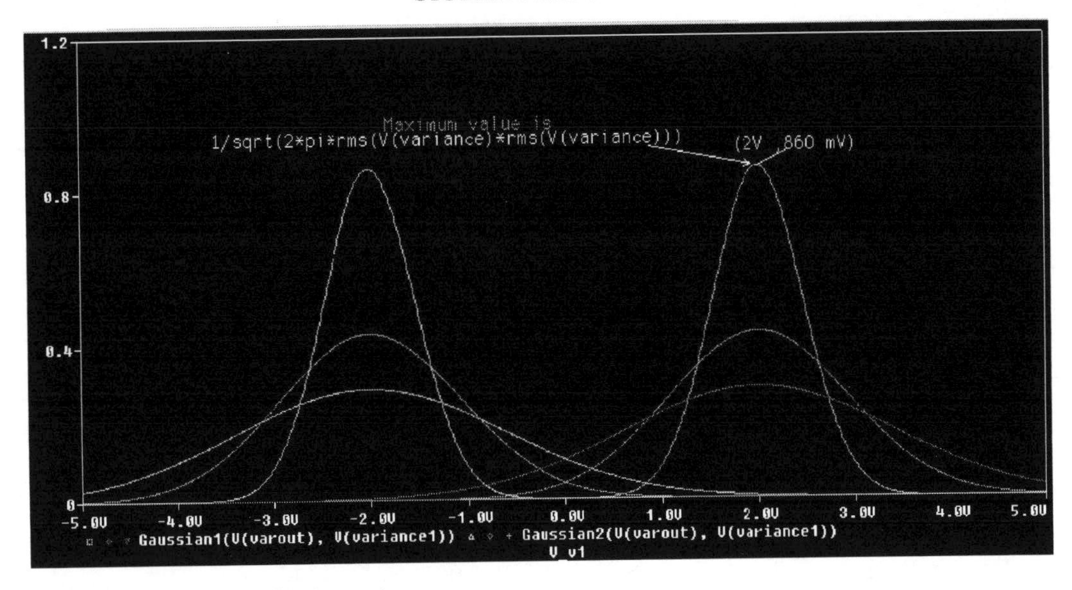

FIGURE 6.30: Increased noise variance.

The maximum value for the Gaussian distribution is 1/sqrt(2*pi*rms(V(variance)*rms(V(variance))). We use the **PARAM** part to define variance and give it a nominal value of 3. Tick **Parametric Sweep** selected from the **Analysis Setup** menu. This enables us to sweep the noise magnitude using the **GAIN** part whose gain is {**variance**}, braces must be included. Select **Global Parameter, Parameter Name = variance, Start** and **End values** are 1 and 3, with an **Increment** of 1. Fig. 6.30 is a plot of the **variance** for three values. The two PDF plots for AGWN are positioned at −2 V and +2 V, and with a midpoint decision threshold of 0 V.

We may plot the two Gaussian distributions in Fig. 6.30 using the @ function to select the last noise value.

To calculate the probability of received data being in the shaded tail areas shown in Fig. 6.29, we integrate the PDF to produce a cumulative distribution function (CDF) as defined in (6.21). Here, V is the signal mean, and is ±2 V. The error function erf(x) is the tail probability area defined as

$$P_e(x) = \text{erf}(x) = \int_V^{\infty} f_x(x)dx = \frac{1}{\sqrt{2\pi\sigma^2}} \int_V^{\infty} e^{\left(\frac{-x^2}{2\sigma^2}\right)} dx. \tag{6.21}$$

Unfortunately, this integral has no closed solution so other techniques must be used to generate the erf(x) values. The probability event is $x = V/\sigma\sqrt{2}$, so the signal to noise ratio for the case of two symbols separated by $2V$ is

$$\text{SNR} = 20\log_{10}\frac{V}{\sigma\sqrt{2}}. \tag{6.22}$$

The total error probability $P_e = P_0 P_{e0} + P_1 P_{e1}$, where P_0 is the base probability of symbol 0 occurrence, P_1 is the base probability of symbol 1 occurrence. P_{e0} is the error probability of symbol 0 and P_{e1} is the error probability of symbol 1. Scrambling the data ensures that $P_0 = P_1 = 0.5$. Therefore, we can write

$$P_e = P_0\int_d^{\infty} f_0(y)dy + P_1\int_{-\infty}^{d} f_1(y)dy. \tag{6.23}$$

The complementary error function erfc(x) is the area under the curve, outside the range $-u$ to $+u$:

$$\text{erfc}(x) = 1 - \text{erf}(x) = 1 - = \frac{1}{\sqrt{2\pi\sigma^2}}\int_V^{\infty} e^{\left(\frac{-x^2}{2\sigma^2}\right)}dx. \tag{6.24}$$

The error function and the complimentary error functions are plotted in **PROBE** using the schematic in Fig. 6.31. The text file is applied using a **VPWL_FILE** generator part. The text file was created in Excel using the in-built function erfc(x). The first column is from $x = 0$ to 4, and the second column is = erf(A1). Fill in the first two rows, and then drag the two rows down the page for the length of the x-variable.

The transient parameters are **Output File Options/Print values in the output file** = 20 µs, **Run to time** = 4 s, **Maximum step size** = (blank). Press **F11** to simulate and produce a plot of erf(x), erfc(x), and the P_e function, as in Fig. 6.32.

$Q(x)$ is the complementary error function, or Q-function, defined in terms of the error function as

$$Q(x) = 0.5\left[1 - \text{erf}\left(\frac{x}{\sqrt{\pi}}\right)\right] = 0.5\text{erfc}\left(\frac{x}{\sqrt{\pi}}\right) \text{ for } x \geq 0. \tag{6.25}$$

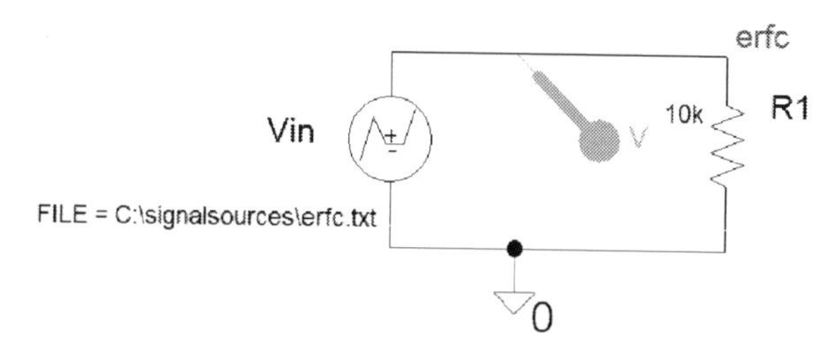

FIGURE 6.31: Plotting erf() and erfc() functions.

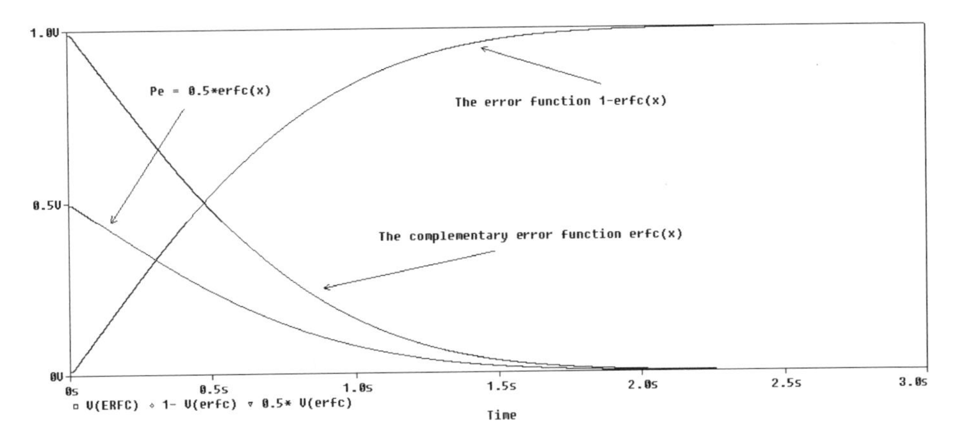

FIGURE 6.32: Error function and complimentary error function and P_e.

6.8.3 Example

An NRZ signal has logic $0 = 0$ V and logic $1 = 5$ V. If the RMS value of the noise is 2.5 V, then the probability of a mistaken symbol is $Q(\text{SNR}/2) = Q(5/2 \times 2.5) = 0.159$, or 16% of bits are identified in error. If the noise is reduced to 1 V, then the probability of an error occurring is $Q(5/2 \times 1) = Q(2.5) = 0.006$, i.e., 6 pulses in 1000.

6.9 BIT ERROR RATE (BER)

Noise and a limited communication channel bandwidth causes errors to be introduced by the receiver because it cannot distinguish between a noise signal and the data signal. Errors are quantified by defining the probability of an error occurring called the bit error rate (BER), and is the number of errors per million bits sent, i.e., one error in one million, $\text{BER} = 10^{-6}$. BER ranges from a poor value of 10^{-3} to a good value of 10^{-9} and is a function of the signal to noise ratio (SNR). It is normal to redefine the SNR for digital signals as an $E_b N_0$ ratio. Consider a bit rate of R bps in a channel with bandwidth B Hz, and in the presence of white noise with power density N_0 W/Hz. If the received signal is S W signal power, with each bit taking $1/R$ s to transmit, then the energy per bit is $E_b = S/R$ W s (or joules). The signal to noise power ratio in a channel with total noise power $N_0 B$ W is

$$\text{SNR} = \frac{E_b}{N_0}\frac{R}{B} = \frac{E_b}{N_0} * [\text{bandwidth efficiency}]. \tag{6.26}$$

The bandwidth efficiency (spectral efficiency) is R/B. A system SNR, for a given bandwidth efficiency, is proportional to (E_b/N_0), and hence the BER. The channel capacity for a

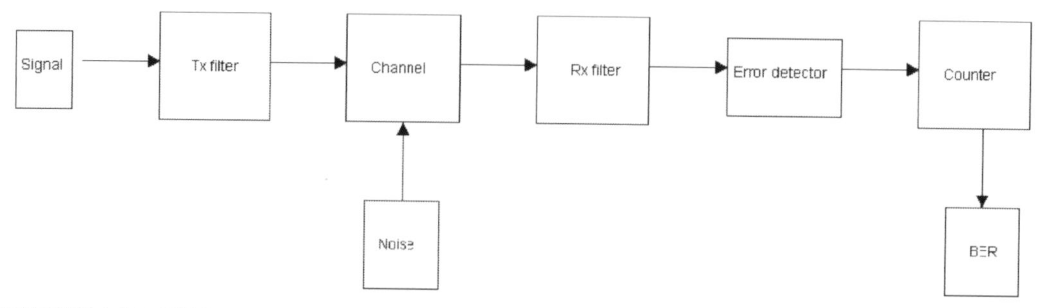

FIGURE 6.33: BER measurement.

multilevel digital channel is

$$C = W\log_2\left\{1 + \frac{E_b R}{N_0 B}\right\}. \qquad (6.27)$$

A symbol error rate, or SER, also defines the error rate for symbols in M-ary signaling. We may plot the BER using the schematic shown in Fig. 6.33.

Block 1 contains the NRZ/NRZB data signal source. The transmission filter in block 2 contains a filter to modify the spectrum before transmission. A pseudorandom additive noise source, created in Matlab, simulates the noise that attaches itself to the signal during transmission and noise contributed by the receiver. We vary the noise amplitude to simulate different signal to noise ratio channel conditions. The error detector generates a pulse for every error that occurs. These errors are then counted in the next block shown in Fig. 6.34, and the BER then estimated. Change the noise level and recalculate the SNR: The output from the bit error meter is shown in Fig. 6.35. Count the numbers of errors from the left and place a cursor to verify the error number.

$$\text{SNR} = 20\log\frac{V_s}{V_n} + 10\frac{2B}{R}. \qquad (6.28)$$

Here V is the RMS of the received signal, and V_n is the RMS noise voltage. B_n is the noise bandwidth, and R is the bit rate.

The bit error rate is the number of bits received in error for the total number of transmitted bits. For example, a BER of 10^{-6} is one error bit for every million bits transmitted.

6.10 CHANNEL CAPACITY

In Fig. 6.36, place an **ABM** part and insert Eq. (6.27) in the **PARAM** part. To convert from base 2 to base 10, consider the following: $\log_2(x) = y$, so that $2^y = x$. Applying logs to both sides gives $y\log_{10}(2) = \log_{10}(x)$, therefore $y = \log_{10}(x)/\log_{10}(2) = 3.32\log_{10}(x)$.

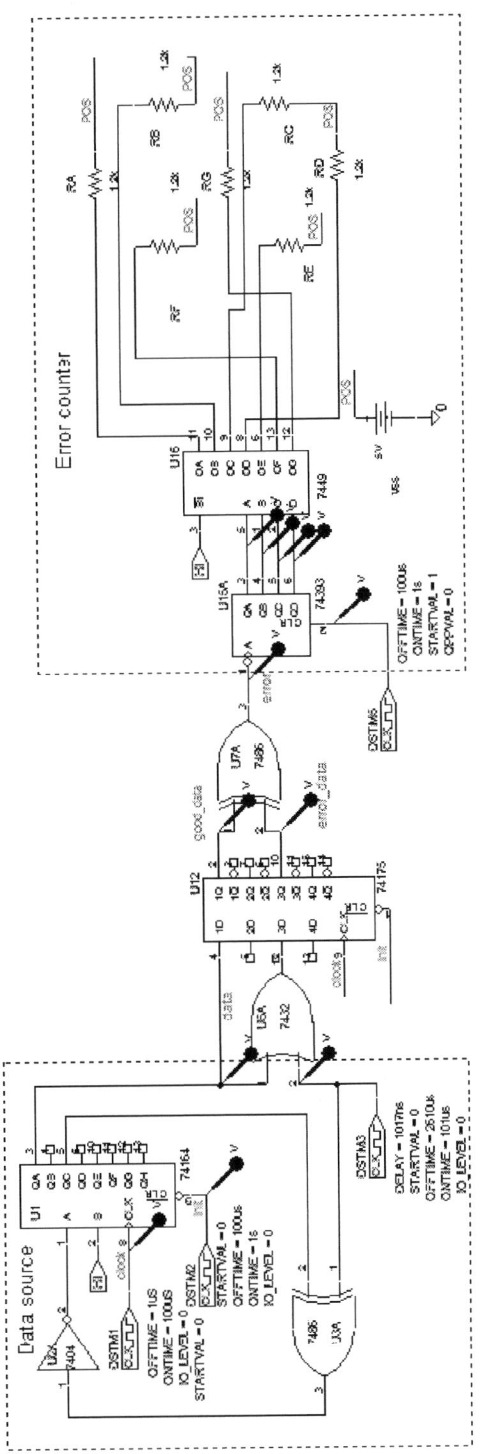

FIGURE 6.34: BER measurement instrument.

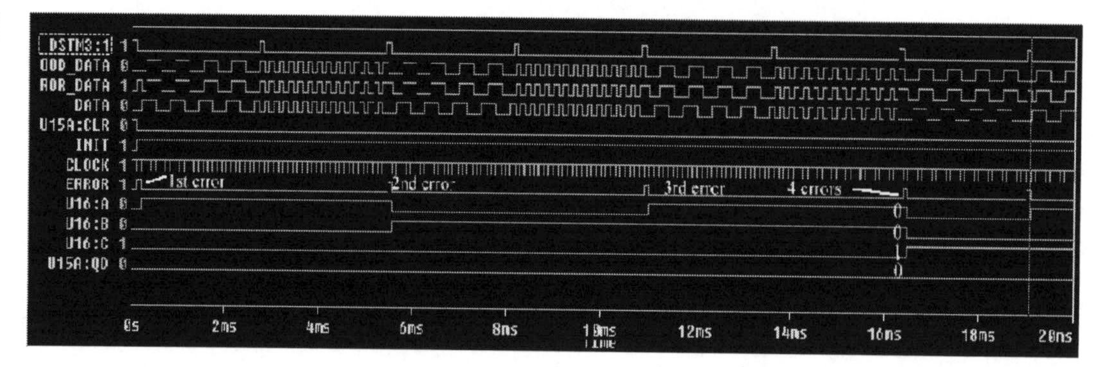

FIGURE 6.35: Output from bit error meter.

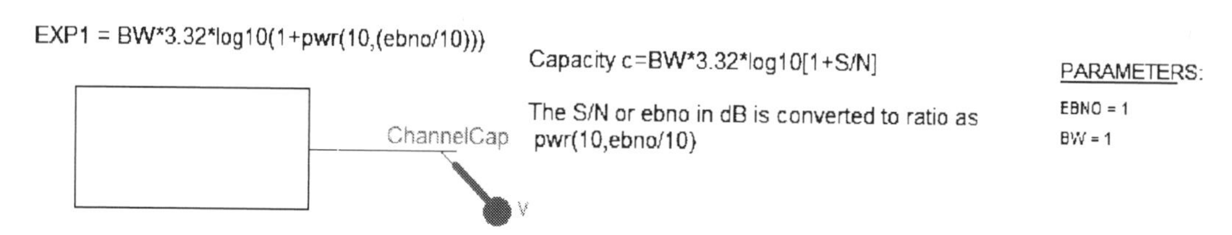

FIGURE 6.36: Channel capacity.

To investigate channel capacity for different E_b/N_0 ratios, select **Analysis Type: DC Sweep, Sweep Variable Global Parameter, Name** = ebno, **Start Value** = 5, **End Value** = 20, **Point/Decade** = 0.1, **linear**. From the **Analysis/ Parametric** menu, set the bandwidth parameters **Global Parameter Name** = BW, **Start Value** = 300, **End Value** = 3500, **Increment** = 500, **linear**. The channel capacity in bits per second is displayed in Fig. 6.37. The **PROBE** y-axis is then calibrated in bps. After simulation, tick OK after swept **Available Sections** and select **OK**.

6.10.1 Channel Capacity for Different M-ary Levels

In this experiment, the number of M levels is varied over a range of values in order to investigate the effect on channel capacity. Draw the schematic in Fig. 6.38.

To vary the number of levels M, set the **DC Sweep** parameters: **Global parameter Name** = M, **Start Value** = 2, **End Value** = 32, **Increment** = 2, **linear**. From the **Analysis/ Parametric** menu, set the bandwidth parameters **Global parameter Name** = BW **Start Value** = 1k, **End Value** = 64k, **Increment** = 8k, **linear**. Fig. 6.39 illustrates how the channel capacity changes as the number of levels and the bandwidth are varied. The y-axis should be in units of kb/s and not kV, but PSpice does not permit changing the y-axis units.

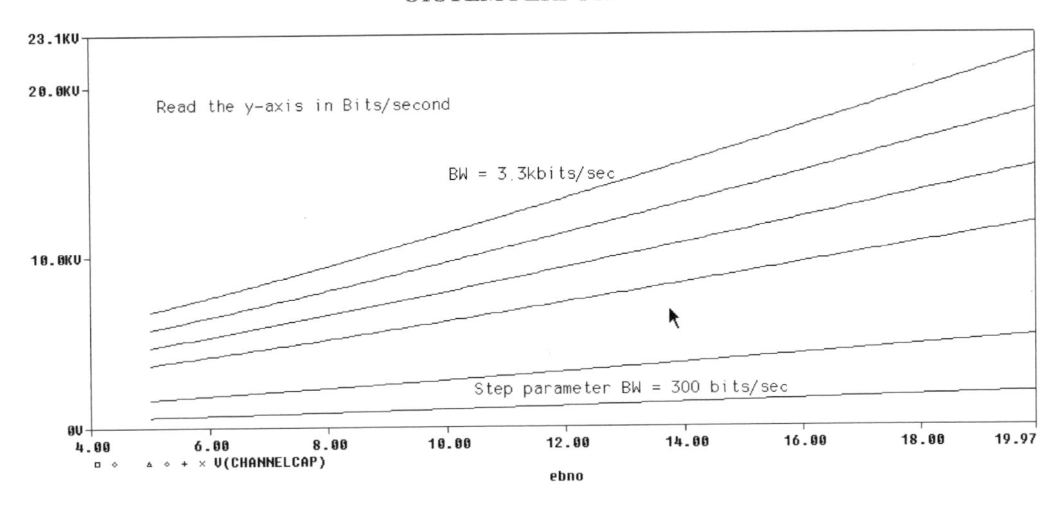

FIGURE 6.37: Channel capacity versus E_b/N_0 ratio.

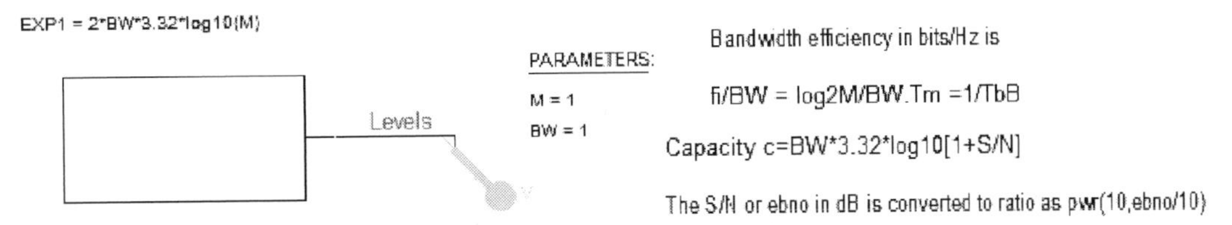

FIGURE 6.38: Channel capacity versus M.

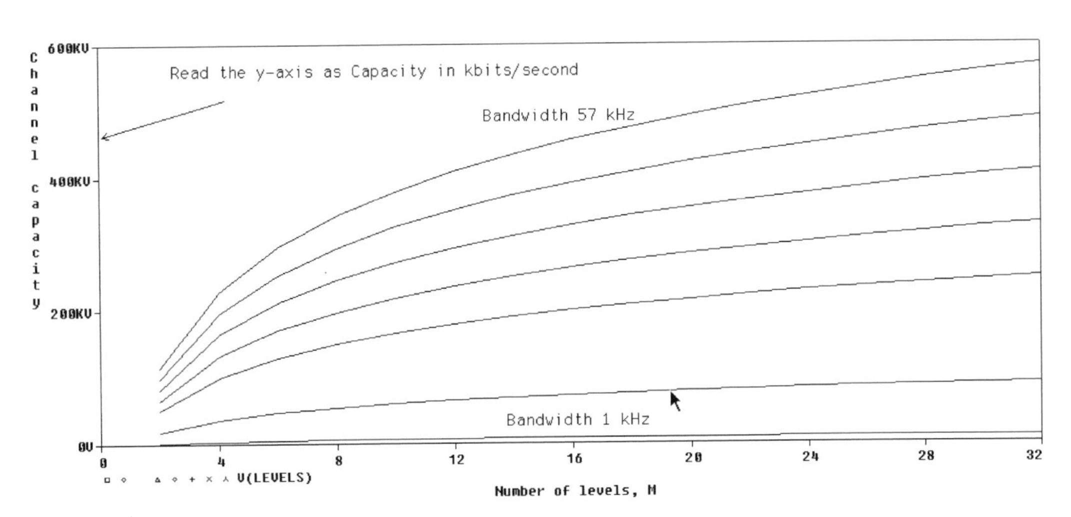

FIGURE 6.39: Channel capacity versus M for different BW values.

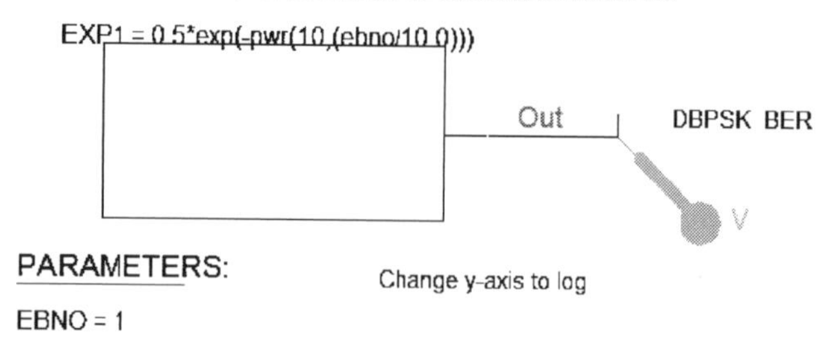

FIGURE 6.40: Investigate BER versus E_b/N_0 ratio.

6.10.2 BER Performance for a Range of E_b/N_0 Ratios

Draw the circuit in Fig. 6.40.

$$P_e = 0.5 * [1 - \text{erf}(\sqrt{10^{(\frac{\text{ebno}}{10})}})]. \tag{6.29}$$

This formula uses the ebno ratio by taking the antilog of the dB ratio as $10^{\text{ebno}/10}$:

$$\text{BER} = \frac{0.5}{\sqrt{(E_b/N_0)\pi}} e^{-E_b/N_0}. \tag{6.30}$$

Press **F11** to simulate. From **PROBE** select y-axis and change **Data Range** to **User Defined**. Set the range from 1 μV to 150 mV and **Logarithmic** as shown in Fig. 6.41.

The BER for different systems is investigated using the schematic in Fig. 6.42. Here the BER for each system is defined in a unique expression.

After simulation, change the y-axis data range from 0 to 0.1p, and scale from **Linear** to **Log**. Communication textbooks will normally show the y-axis as 10^{-1}, 10^{-2}, etc., but this

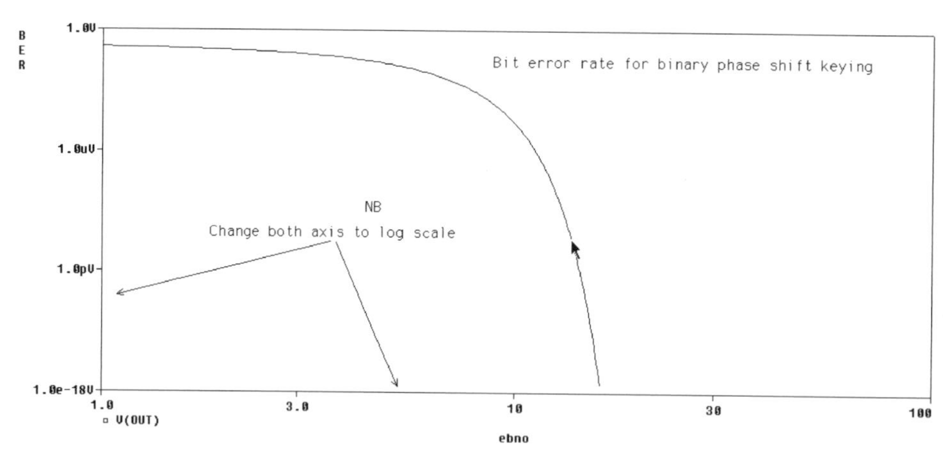

FIGURE 6.41: BER for BPSK.

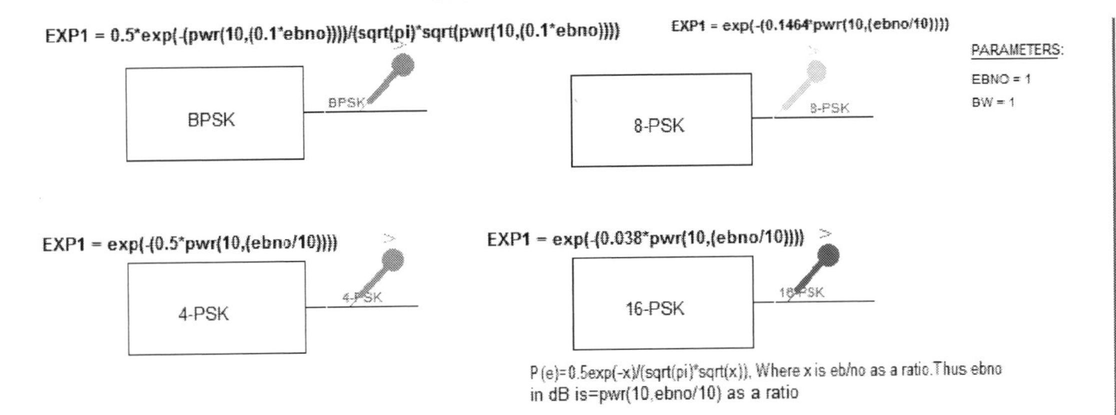

FIGURE 6.42: *M*-ary PSK BER comparison.

format is not possible in **PROBE,** so we must change the y-axis to a log plot as shown in Fig. 6.43

For multilevel modulated carrier systems, we plot the BER as

$$P_e = e^{-(C/N)\sin^2(\pi/N_\phi)}, \tag{6.31}$$

where C/N is the carrier to noise ratio, and N_ϕ represents the phase modulation scheme, i.e., the number of phases. For binary PSK, $N_\phi = 2$. Plot the BER using the schematic in Fig. 6.44.

Select **Analysis** and tick **Parametric Sweep**. Select **Global Parameter** and **Parameter Name** = CN. In the **Value** box enter $N = 2$, 4, 8, 16, and 32-PSK to plot the C/N for different values of CN as plotted in Fig. 6.45.

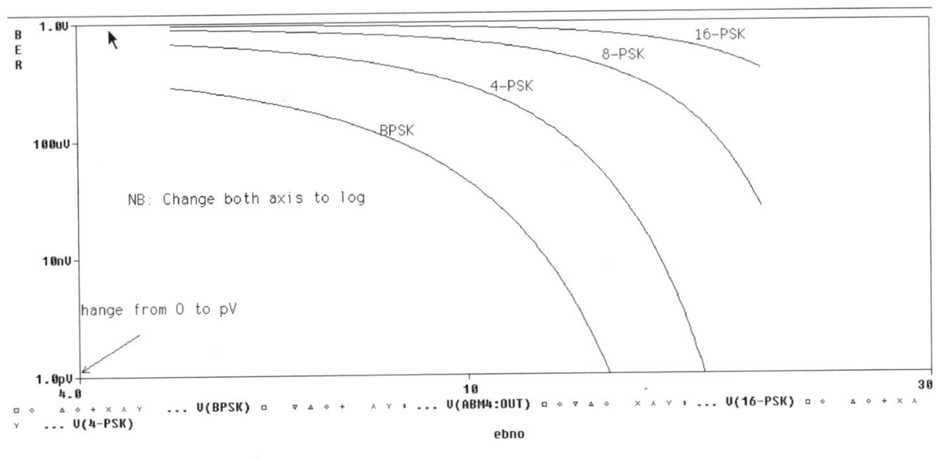

FIGURE 6.43: BER comparison.

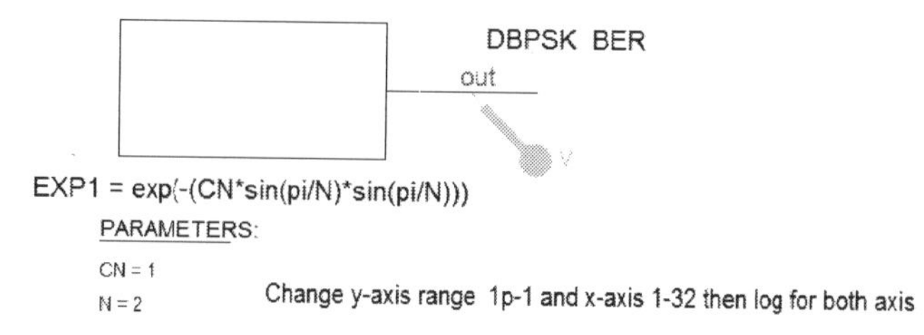

FIGURE 6.44: BER as a function of C/N.

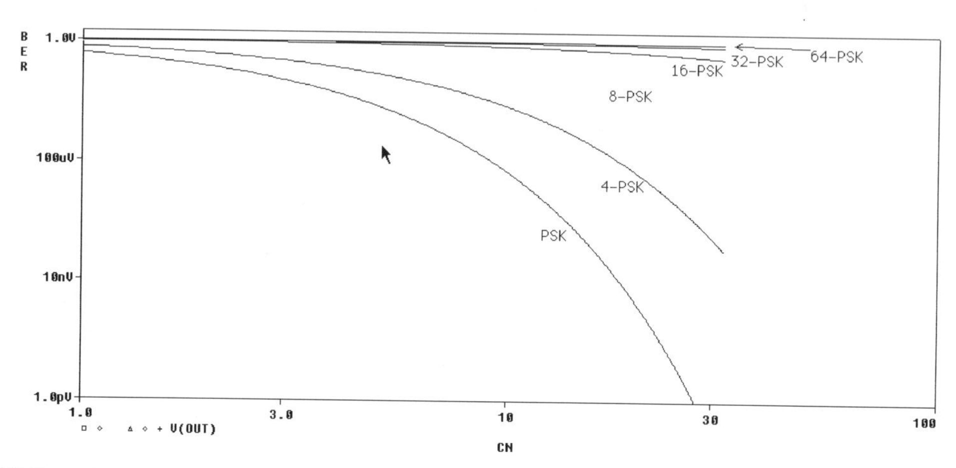

FIGURE 6.45: BER for M-ary systems.

6.11 CYCLIC REDUNDANCY CHECK

Cyclic redundancy check (CRC) is an error detection and correction encoding technique. The total message block in CRC, including the check bits, is a polynomial defined as

$$p(x) = x^n + x^{n-1} + \cdots + x^2 + x + 1. \tag{6.32}$$

The highest order bit transmitted is x^n. For example, the binary word 10011 is

$$x^4 + x + 1. \tag{6.33}$$

The degree of the polynomial is one less than the number of encoding bits and the original message $M(x)$ contains m bits, where $G(x)$ is the generator polynomial, $T(x)$ is the transmitted encoded polynomial, and $E(x)$ is error polynomial. The encoded message $T(x)$, $M(x)$, is shifted by the number of added check bits c to the left, and is the degree of the generator represented as $M(x)x^c$. The shifted message is divided by $G(x)$, leaving a remainder that is added to the shifted message to produce $T(x)$. This polynomial can always be divided

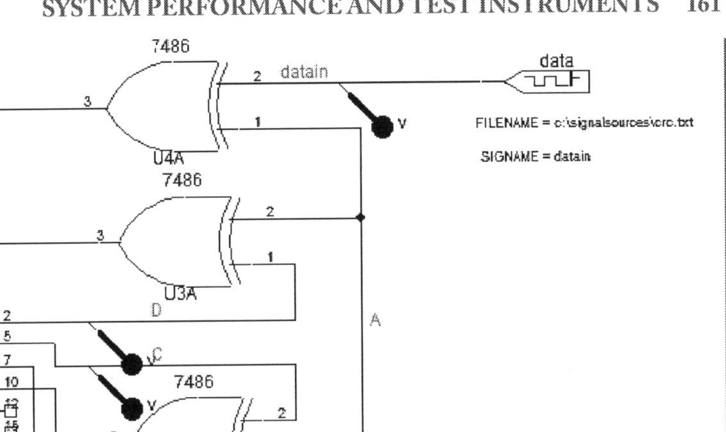

FIGURE 6.46: CRC generator.

by $G(x)$ without a remainder:

$$\frac{M(x)x^c}{G(x)} = Q(x) + \frac{R(x)}{G(x)} \Rightarrow M(x)x^c = Q(x)G(x) + R(x). \tag{6.34}$$

$Q(x)$ is the quotient and $R(x)$ is the remainder of degree one less than $G(x)$.

$$T(x) = M(x)x^c + R(x) = Q(x)G(x) \tag{6.35}$$

Addition and subtraction are the same in modulo-2 arithmetic. Therefore, $T(x)$ is exactly divisible by $G(x)$ giving $Q(x)$. Draw the schematic shown in Fig. 6.46.

Verify that the required outputs shown in Fig. 6.47 are obtained.

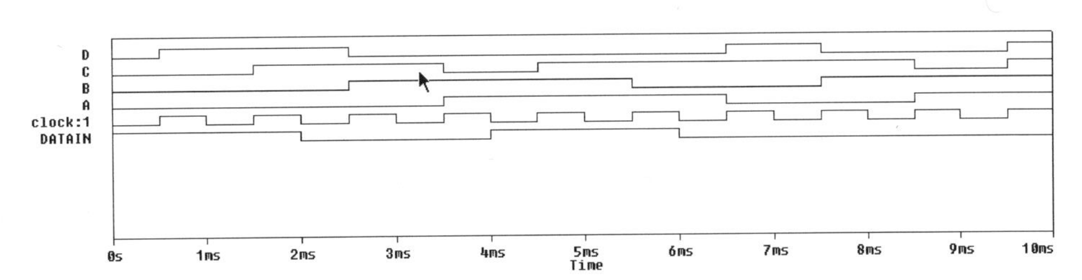

FIGURE 6.47: CRC generator waveforms.

6.12 EXERCISES

1. Investigate SNR using the band-limited noise production in Fig. 6.48.
2. Apply the jitter generator to the schematic in Fig. 6.12 and investigate the effect on the timing parameters using the eye diagram.
3. Replace the AND gate with an XOR gate in Fig. 6.17 and investigate.
4. Investigate a carrier recovery schematic that bandpass limits the spectrum, and hence the noise, before it is applied to a squaring circuit to remove the modulation. A PLL extracts the doubled frequency and is then frequency divided to give the required carrier signal.
5. Investigate bit error rate (BER) using the incomplete circuit in Fig. 6.34.

   The **FileStim** part applies the digital test signal created with the text editor Notepad in Fig. 6.49.

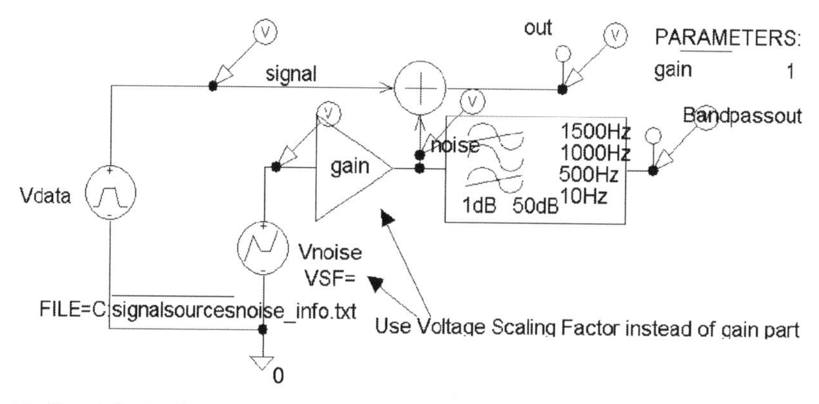

FIGURE 6.48: Band-limited noise.

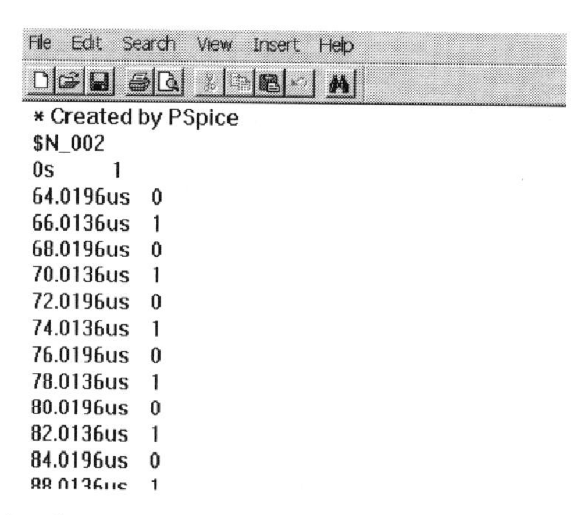

FIGURE 6.49: Test data signal.

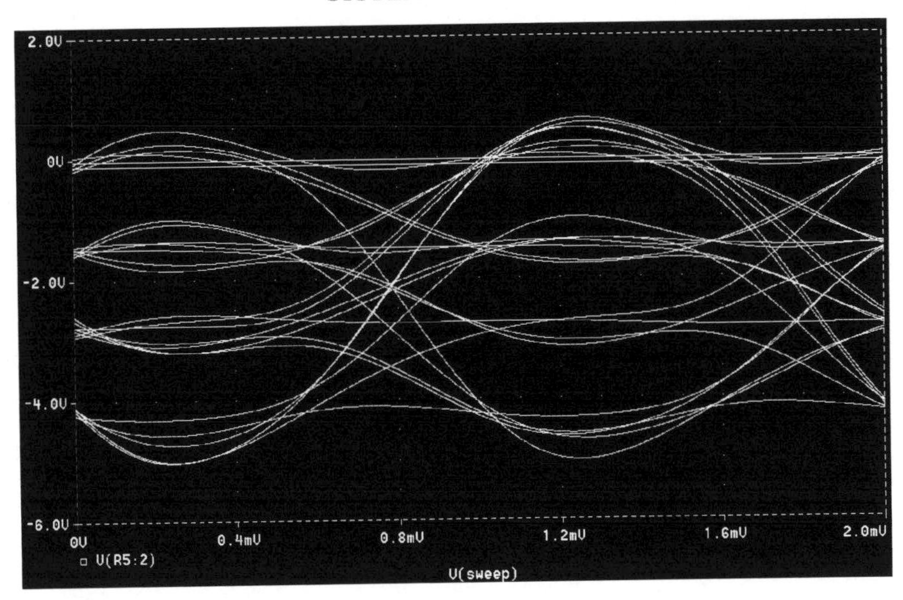

FIGURE 6.50: The eye diagram for a multilevel signal.

The error message "ERROR – Unable to find signal test in stimulus file "C:\PSpice\circuits\see3\testprbs.vec," occurs if you copy the **FileStim** part from another schematic and the header node information %N_002 is different to the node or the wire segment name. Name the **SigName** in the **FileStim** properties to avoid ambiguity problems. The name could be a wire segment name, for example, data.

6. Connect the eye meter to an M-ary system such as 4-PSK to produce the eye diagram shown in Fig. 6.50. Note: after simulation, change the x-axis from time to the swept signal, v(sweep).

   Change the channel bandwidth to produce different eye diagrams.

7. The schematic in Fig. 6.51 shows how a noise file, created in Matlab, is applied with a **VPWL_F_RE_FOREVER** part.

   The transmission line primary line parameters are defined per unit line length (in this case it is per km). The characteristic impedance is calculated at an operating frequency of 2 MHz:

$$Z_0 = \sqrt{\frac{R + j\omega L}{G + j\omega C}} = \sqrt{\frac{(15 + j2\pi 2.10^6 175.10^{-6})}{(9.10^6 + j2\pi 2.10^6 70.10^{-9})}} = 50 - j0.17\ \Omega \simeq 50\ \Omega. \quad (6.36)$$

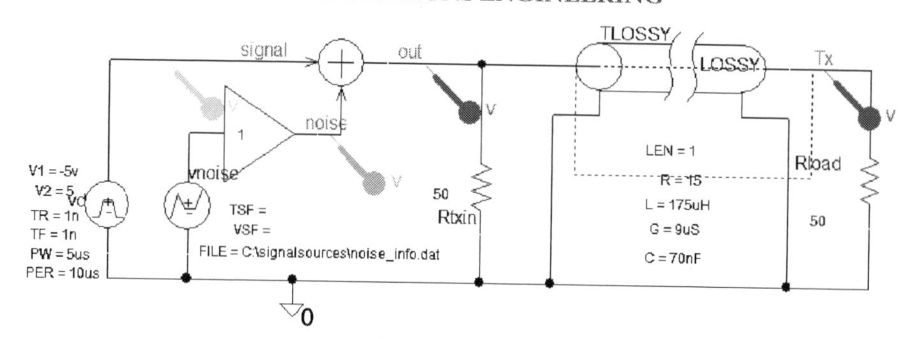

FIGURE 6.51: Signal + noise on a transmission line.

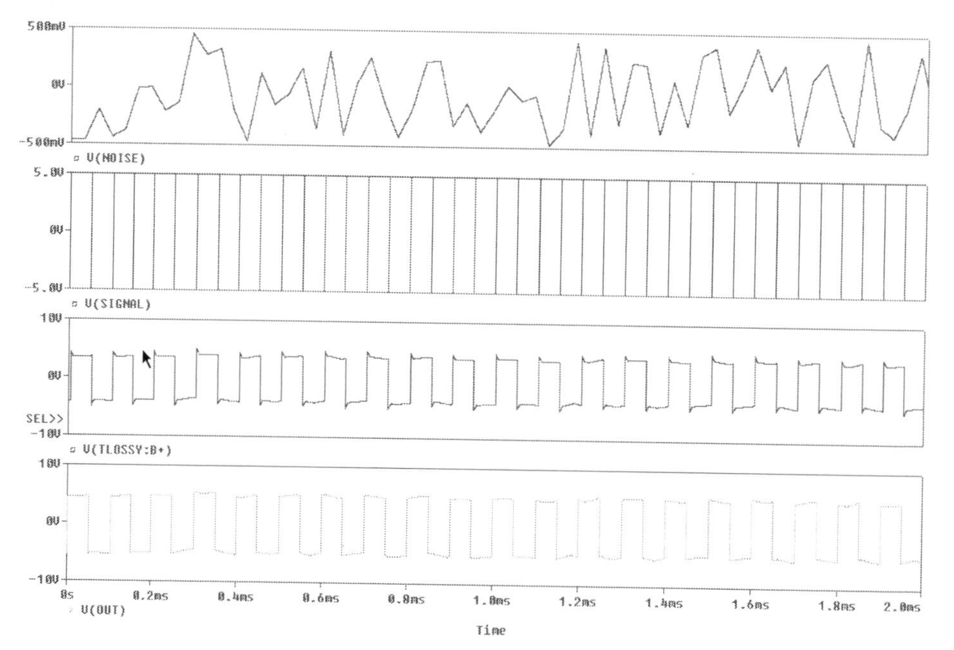

FIGURE 6.52: Time waveforms.

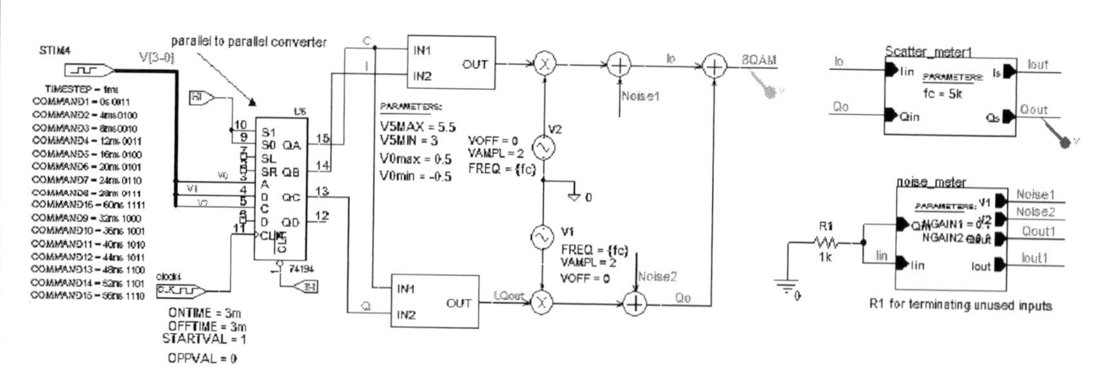

FIGURE 6.53: An 8-QAM system.

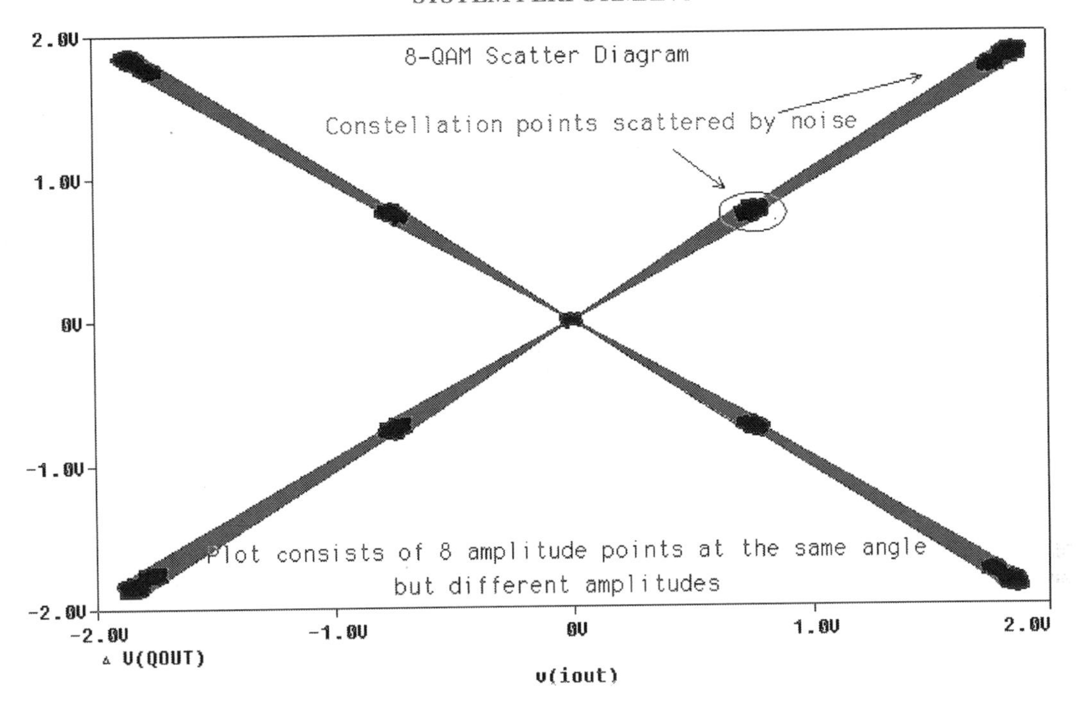

FIGURE 6.54: 8-QAM scatter diagram.

Set the **Analysis** tab to **Analysis Type**: **Time Domain (Transient), Run to time** = 250 μs, and **Maximum Step size** = 1 μs. Press **F11** to simulate and display the signals as shown in Fig. 6.52 [ref 1: Appendix A].

8. Investigate the 8-QAM schematic shown in Fig. 6.35 and plot a scatter diagram. You should get a similar scatter diagram display to that shown in Fig. 6.54.

CHAPTER 7

Direct Sequence Spread Spectrum Systems

7.1 SPREAD SPECTRUM

The actress Hedy Lamarr together with a friend George Antheils, had a patent granted during World War II for a coding system that formed the basis of spread spectrum techniques used in modern communications. In spread-spectrum, the transmitted signal is spread over a much larger bandwidth than the information signal, hence the name "spread-spectrum". Spread spectrum has very good security properties being less susceptible to interference compared to other systems, and has many military and civil communication applications including cellular phones and wireless local area wireless networks (WLAN). The two main spread-spectrum techniques used in code division multiple access (CDMA) are: Direct Sequence spread spectrum (DSSS), and Frequency Hopping spread spectrum (FHSS).

DSSS is utilized in the Digitally Enhanced Cordless Telecommunications (DECT) phone system, where a Codec digitizes voice signals to produce a 64 kbit/s (8000×8 bps) data signal. To avoid being overheard by a third party, the system generates a unique code that is extremely difficult to intercept. Each zero and one in the transmitted information signal is given a unique 32-bit code. The signal is then spread over a large bandwidth and transmitted using differential phase shift keying (DPSK), or differential quadrature phase shift keying (DQPSK). A pseudo-random binary sequence (PRBS) (pseudo-random meaning the sequence is not really random but repeats after a certain time) phase-modulates the carrier in a random fashion by a continuous sequence of symbols called chips. The duration of these chips is much shorter than an information bit, so that when each information bit is modulated by the sequence of faster chips it results in a chip rate that is much higher than the original information signal bit rate.

In DSS, the data is multiplied by the PRBS NRZ-B sequence of 5 and -5 values at a frequency much higher than that of the original signal and will spread the information signal over a much wider band. The resulting signal is similar to noise except that the receiver can recover the original data. The basic parts of a DSSS system in Fig. 7.1 comprise a PRBS, the heart of the DSS system, which DIRECTLY modulates the baseband data signal resulting in

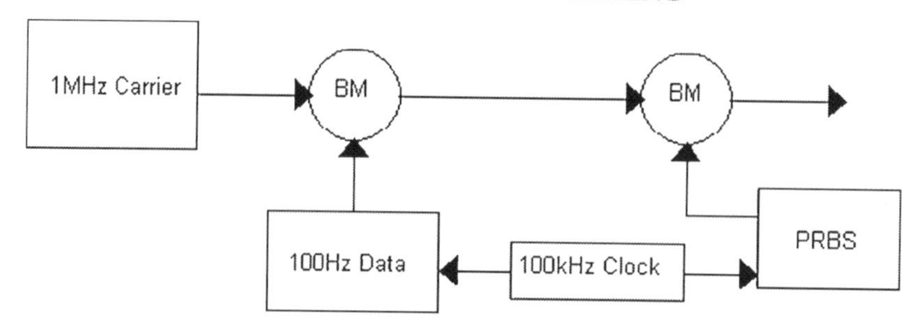

FIGURE 7.1: DSSS system.

the transmitted signal bandwidth being spread by factors up to a 1000. The resultant passband signal spectrum has a sinc-shaped response. An identical PRBS in the receiver demodulates the DSS passband signal, which is then de-spread using an identical carrier.

7.2 PSEUDORANDOM BINARY SEQUENCE PROPERTIES

Since the PRBS is the most important part of the spread spectrum system then some technical details should be examined first. The **balance property** of a PRBS is defined where the number of 1s exceeds the number of 0s by one only. A **run** is an unbroken sequence of a single type of binary digit such as a string of 1s, or a string of 0s and a **run length** is the number of digits contained in a run. In each sequence, half of the runs of each kind are of length 1, a fourth are of length 2, an eighth are of length 3, and it is called the **run property**. An m-sequence contains the following:

- A run of 1s of length m.
- A run of 0s of length $m - 1$.
- A run of 1s and one run of 0s, each of length $m - 2$.
- A run of 1s and two runs of 0s, each of length $m - 3$.
- Four runs of 1s and four runs of 0s, each of length $m - 4$.
- 2^{m-3} runs of 1s and 2^{m-3} runs of 0s, each of length 1.

When simulating digital circuits the flip-flops must be initialized into a certain state. Select the **Simulation Setting** menu and then select the **Options** tab. In the **Category** box select **Gate-Level Simulation**. This will show the **Timing Mode** - select **Typical**. Make sure to select **Initialization all flip-flops to 1**.

Set the Analysis tab to Analysis type: **Time Domain** (Transient), **Run to time** = 200us, and **Maximum step size** = 1us, Press F11 to produce Fig. 7.3.

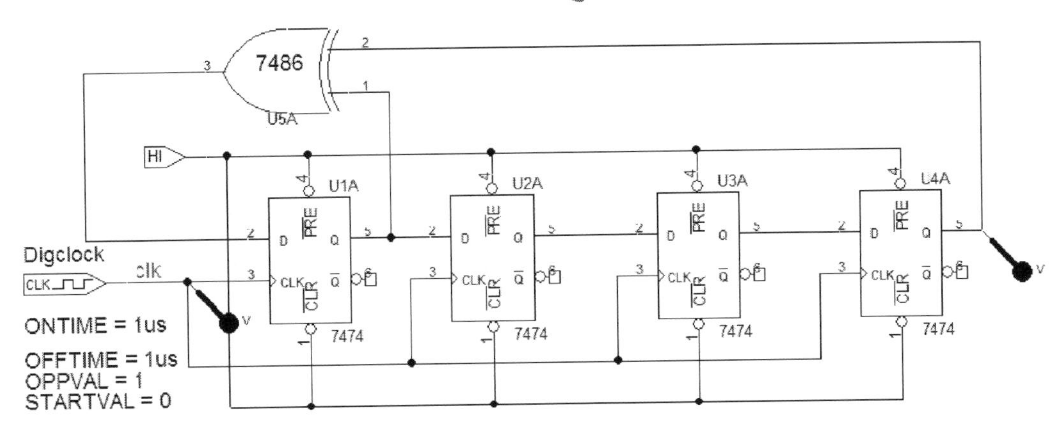

FIGURE 7.2: PRBS generation using D-type flip-flops.

7.2.1 PRBS Generator

The PRBS generator in Fig. 7.2 uses D-type flip-flops connected with feedback paths from m and n bits via an exclusive OR gate (XOR). A **DigClock** part clocks the sequence that repeats after N clock pulses and has a maximum sequence length $N = 2^m - 1$. Most taps generate recursive sequences but only certain feedback combinations produce maximal-length m-sequences.

The taps to produce different sequence lengths are shown in Table 7.1.

A longer PRB sequence is generated by replacing the flip-flops with 8-bit, parallel-out serial-in, shift registers, as shown in Fig. 7.4. Feedback via a 7486 XOR gate connected across selected outputs generates a 32-bit sequence as shown in Fig. 7.3.

7.2.2 Vector Part

To record and log the PRBS, we attach a **VECTOR** part to the output as shown in Fig. 7.4. This part records the ASCII digital signal at this point and logs it as C:\Pspice\Circuits\signalsources\data\prbs.txt. Select the **VECTOR** part from the library, and, after placing it at the desired digital signal output, type in the file name and where it is to be located. The **VECTOR** part attached to pin 11 of U1 records the digital signal at this point

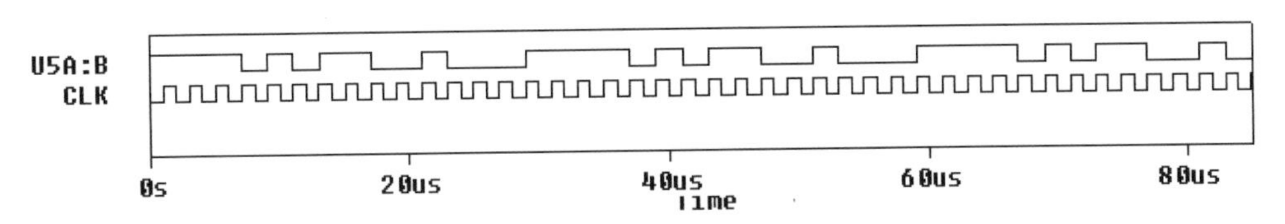

FIGURE 7.3: PRBS and clock signals.

TABLE 7.1: Taps Versus Sequence Length

STAGES m	CODE LENGTH L	TAPS
6	63	(1,6)
17	131072	(1,15)
32	4.2950×10^9	(1,11) (31,32)
33	8.5899×10^9	(1,21)

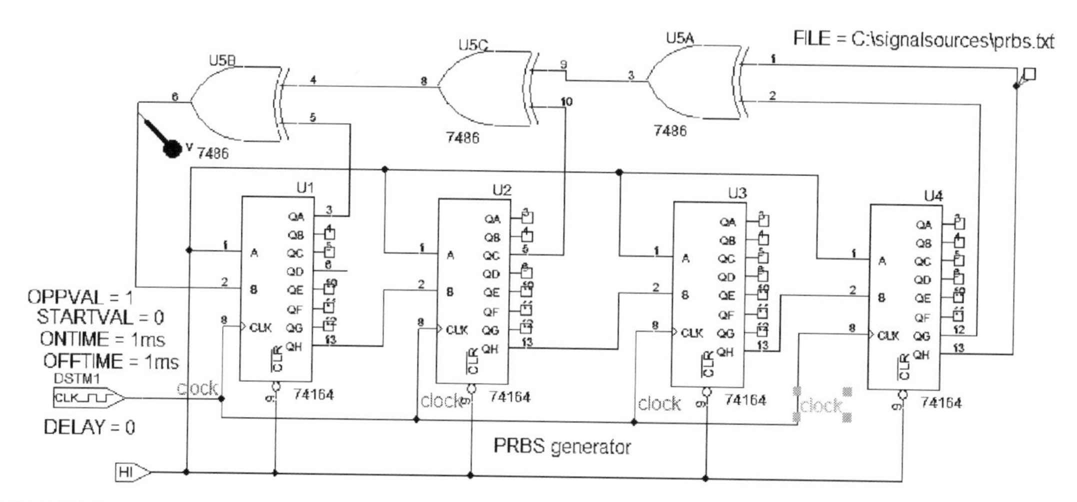

FIGURE 7.4: PRBS generator.

and the file created is used as an input signal in the DSS receiver considered shortly. This is especially useful where a circuit is large and exceeds the evaluation restrictions. To analyze this system, it is best to use the restricted-length PRBS generator in Fig. 7.5. The **DigClock** part clock has equal on and off times.

Set the Analysis tab to Analysis type: **Time Domain** (Transient), **Run to time** = 100ms, and **Maximum step size** = (left blank), press F11 to produce Fig. 7.6.

The m stage serial shift register has a maximum sequence length defined as

$$N = 2^m - 1. \tag{7.1}$$

Here N is the number of chips, where a chip is the time it takes to transmit a bit (a single symbol) of a PRBS code. The m-sequence code length period in seconds, for a clock

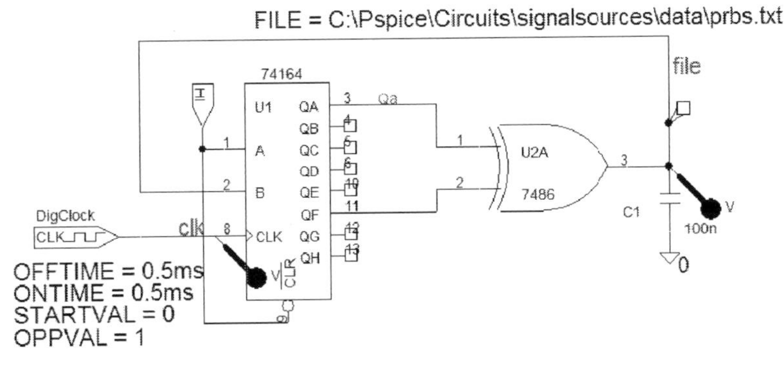

FIGURE 7.5: Simple PRBS generator.

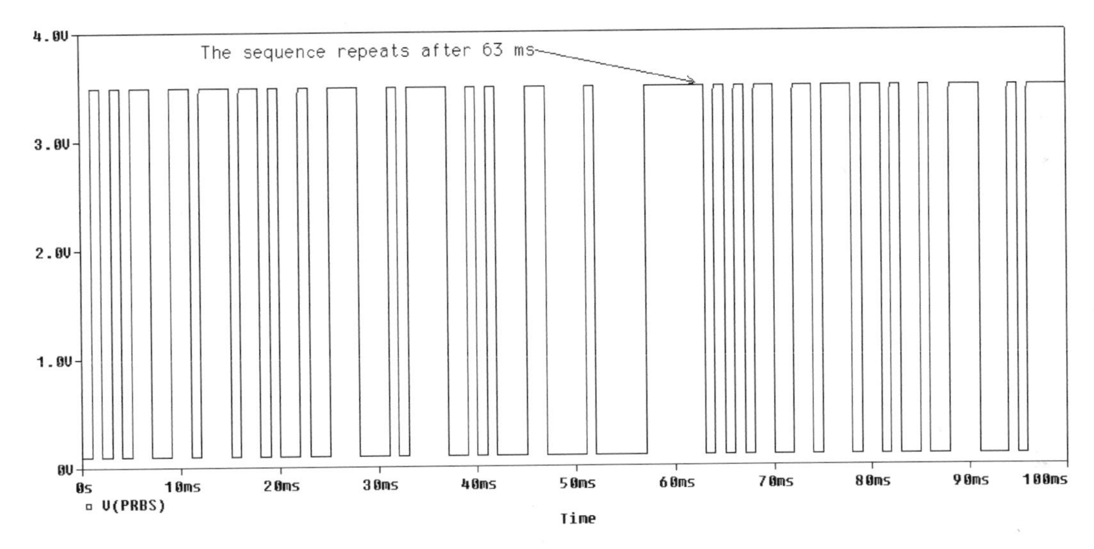

FIGURE 7.6: The PRBS signal.

$f_c = 1/T_c$, is

$$T_{\mathrm{DS}} = NT_c. \tag{7.2}$$

We may use this equation to calculate the PRBS length. Rearranging (7.2) (invert the time parameters) as

$$R_{\mathrm{DS}} = \frac{R_c}{N} \tag{7.3}$$

feedback points 1 and 6 produce a sequence length $N = 2^6 - 1 = 63$, as measured in Fig. 7.6. A capacitor connected across the output suppresses overshoot, but is also required in order to display the **FFT** of a digital signal. However, don't forget to decrease the capacitance value if you

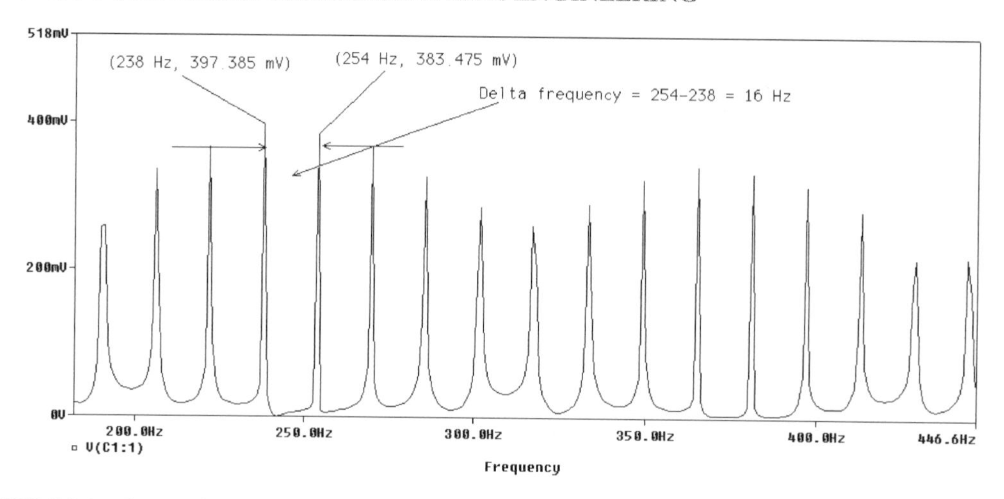

FIGURE 7.7: Spectral spacing.

increase the clock frequency. Set the Analysis tab to Analysis type: **Time Domain** (Transient), **Run to time** = 1s, and **Maximum step size** = 10us, Press F11 and press the FFT icon when the PRBS appears. The length of the **PRBS** is determined indirectly from the frequency spectrum shown in Fig. 7.7.

Applying the magnifying tool to a section of the frequency spectrum allows you to measure the spacing between spectral lines as 15.74 Hz. The relationship between N, T_c, and f (frequency between individual spectral lines) is

$$f = \frac{1}{NT_c} \Rightarrow N = \frac{1}{fT_c} = \frac{1}{f(1/f_c)} = \frac{f_c}{f}. \tag{7.4}$$

The sequence length calculated from the 15.74 Hz spectral spacing (for a 1 kHz clock) is

$$N = \frac{f_c}{f} = \frac{10^3}{15.74} = 63. \tag{7.5}$$

Note that the all-zero state is where all zeros are circulated and is prevented from occurring by setting the initial conditions to one using a **Hi** part, or by using an initialization pulse obtained from an LPF *CR* network.

7.2.3 PRBS Applications

Fig. 7.8 shows a 555 timer IC clocking the 33-stage PRBS. The IC clear signal is produced from a C_3R_3 network connected to a DC supply (alternatively, you may use a **Hi** part). The initially uncharged capacitor charges up to 5 V in approximately five time constants. Be sure that the **Skip initial transient solution** in the **Analysis** menu/**Transient** submenu is not ticked.

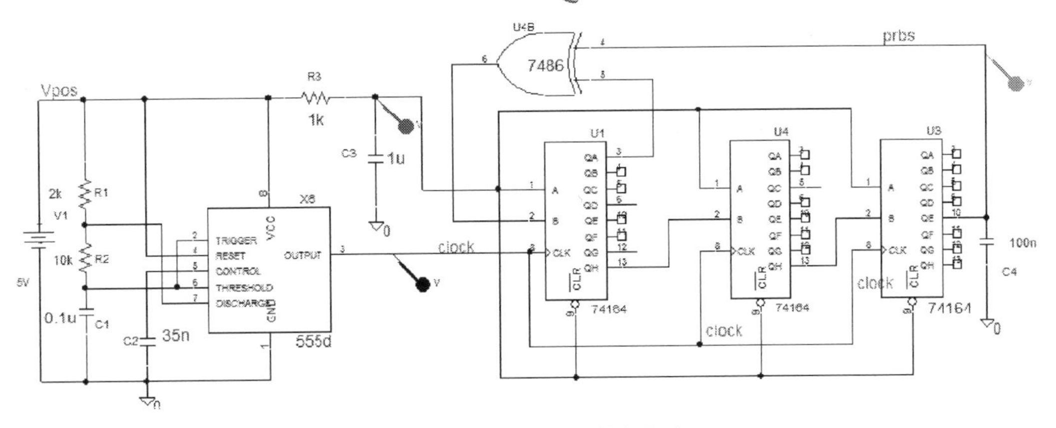

FIGURE 7.8: A buffered output 33-stage PRBS with a 555 clock.

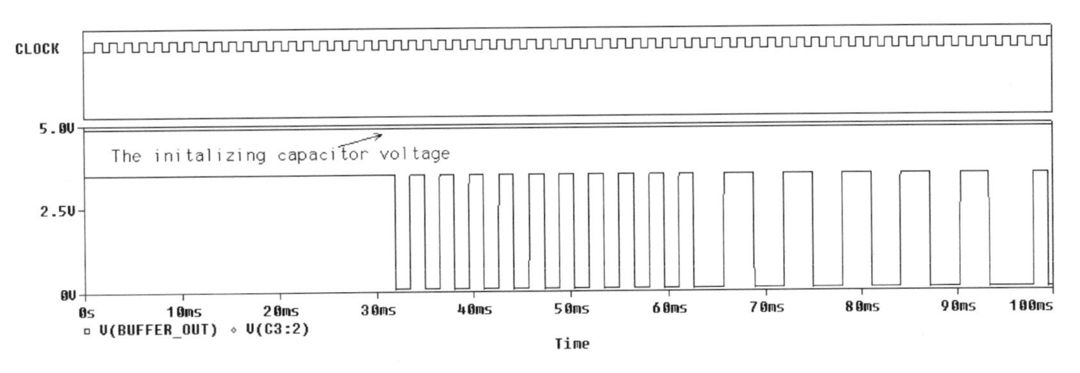

FIGURE 7.9: PRBS output overshoot suppression.

From the **Analysis Setup**, select **Digital Setup** and set the **Initialize all flip flops** to **All 1**. A 100 nF capacitor placed across the PRBS output before the unity gain buffer amplifier reduces overshoot transients. Set the Analysis tab to Analysis type: **Time Domain** (Transient), **Run to time** = 100ms, and **Maximum step size** = (1us), Press F11 to simulate and produce Fig. 7.9 which shows the system clock and PRBS signals.

The initializing voltage generated by $R3 + C3$ is shown but because the **Skip initial transient** is not ticked we do not see the charging up phase.

7.3 DIRECT SEQUENCE SPREAD SPECTRUM TRANSMITTER

Spread spectrum (SS) signals have noise-like properties and are transmitted at a low spectral power density (W/Hz) making them hard to intercept. The low signal power density gives these signals an advantage because narrow band signals may occupy the same band with little

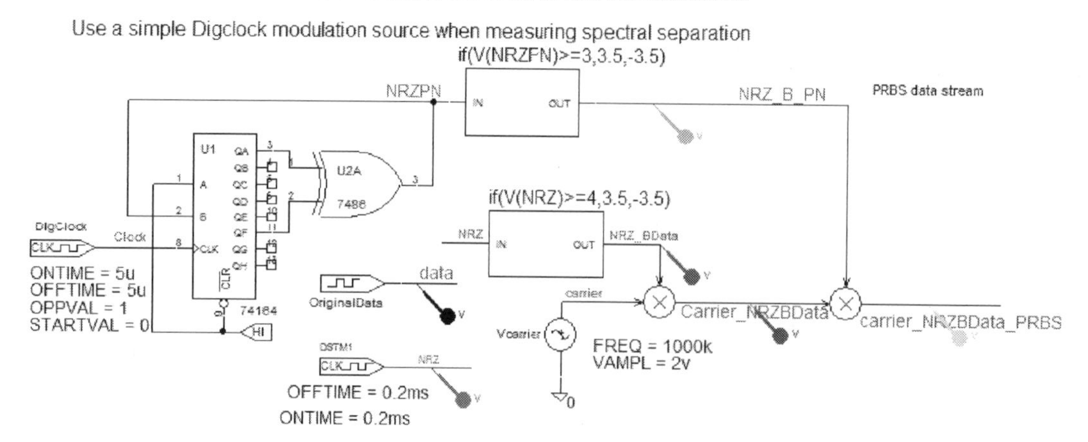

FIGURE 7.10: Spread spectrum transmitter.

or no interference between them. Fig. 7.10 shows a transmitter circuit consisting of a simple PRBS signal generator and **ABM** multipliers. We need to convert the PRBS and modulating data information to NRZ-B format. We do this simply by using **ABM1** parts with the **IF then else** construct as if (V(NRZ) > =3, 3.5, −3.5). What this does is to convert the voltage NRZ 0-5 volts on the input wire segment called NRZ, to −3.5 V and 3.5 V.

An alternative to the BM is the four-quadrant (meaning it can handle all input signal sign permutations) AD633 multiplier manufactured by Analog Devices.4. Another multiplier is the Motorola multiplier IC MC1496, but this IC requires extra external components, and unsymmetrical dual power supply voltages.

7.3.1 STIM Generator Part

To investigate SS principles, it is best to use a PRBS with a short repeating sequence, or a signal data from a simple square wave **VPULSE** generator part. For a more authentic random signal source, use a **STIM1** part whose parameters are set as shown in Fig. 7.11. The plus sign before the time increment shows it is *relative* time and builds on the previous signal transient. Without the plus sign, you must enter the actual time at each increment, i.e., 0.1ms, 0.2 ms, etc.

The GOTO command makes the sequence repeat a number of times, whereas GOTO 2-1 makes it repeat indefinitely. Use a **VSIN** part called Vcarrier as the carrier parameters. Note that the carrier frequency is much higher than the data rate and appears as a green band.

Set the Analysis tab to Analysis type: **Time Domain** (Transient), **Run to time** = 10ms, and **Maximum step size** = (left blank), Press F11 to produce Fig. 7.12.

The DSSS waveforms are shown in Fig. 7.13.

COMMAND1	LABEL=STARTLOOP
COMMAND2	+0ms 1
COMMAND3	+.2001ms 1
COMMAND4	+0.2ms 0
COMMAND5	+0.2ms 0
COMMAND6	+0.2ms 1
COMMAND7	+0.2ms 0
COMMAND8	+0.2ms 1
COMMAND9	+0.2ms 1
COMMAND10	+0.2ms 1
COMMAND11	+0ms GOTO STARTLOOP 2
TIMESTEP	+1ms
Value	STIM1
WIDTH	

FIGURE 7.11: Data signal parameters.

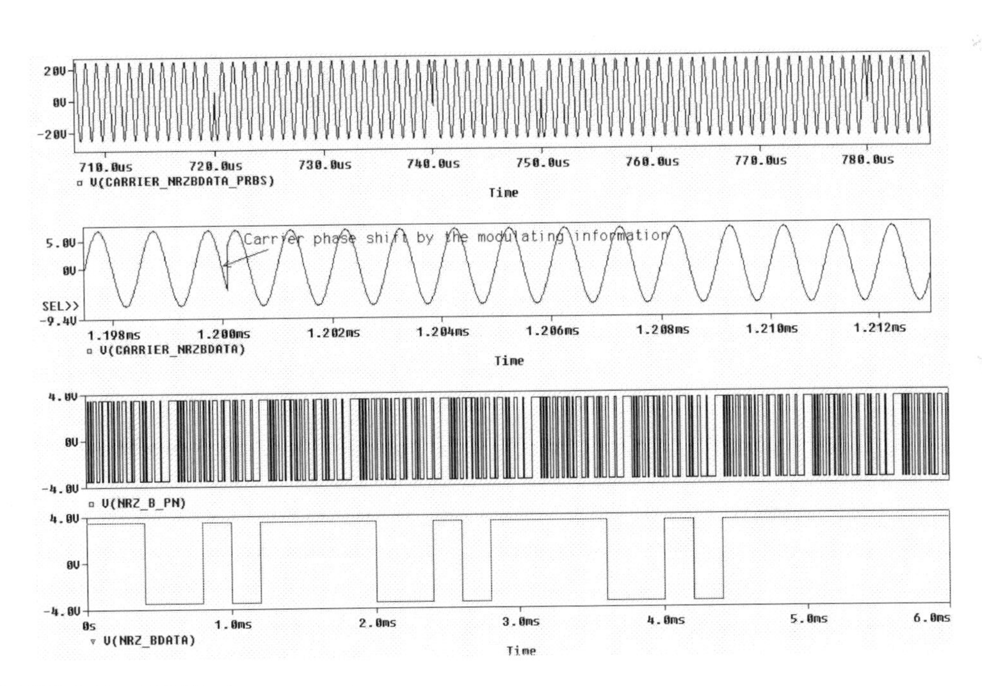

FIGURE 7.12: 100 Hz data parameter.

Apply the magnifying tool to a section of the waveform and examine the signal as displayed in Fig. 7.14. The DSSS spectrum shows the data signal shifted and using the magnifying tool we can examine the spreading effect of signal in detail. Compare the resultant spectrum before and after multiplying the modulated carrier by the PRBS sequence. From the magnified spectrum we can measure the spectral line separation [ref 4: Appendix A].

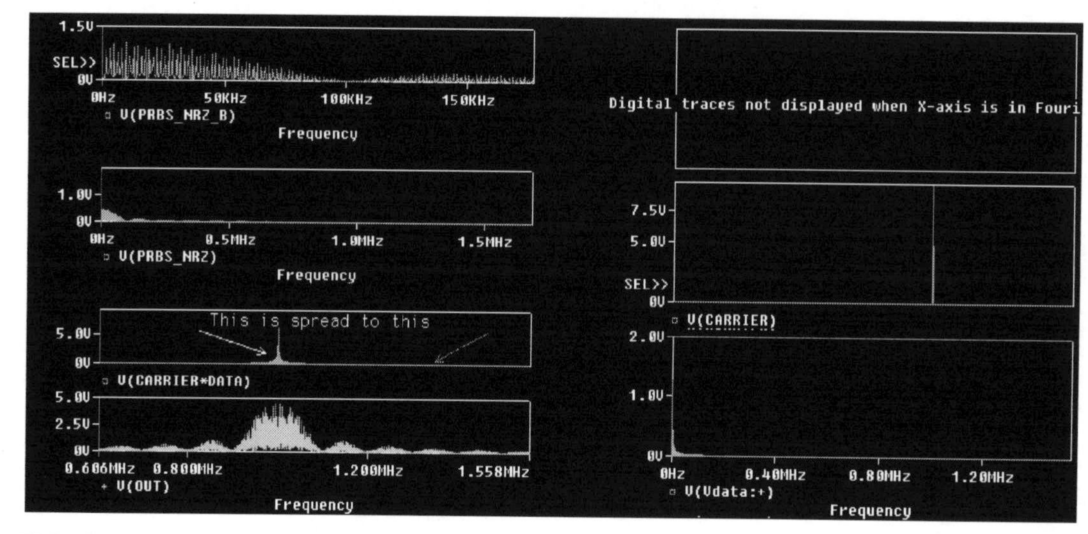

FIGURE 7.13: DSSS time signals.

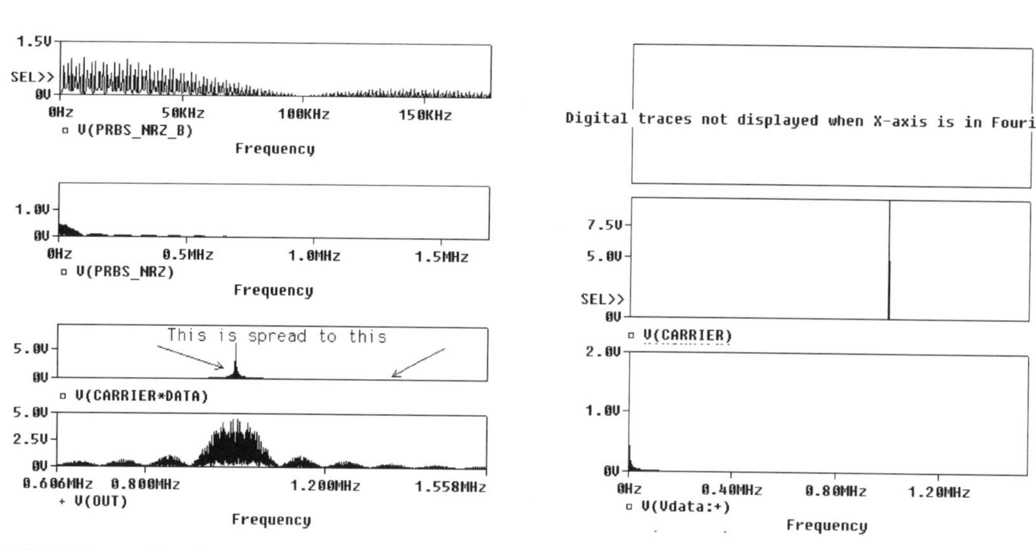

FIGURE 7.14: Further magnification.

Modulating an RF carrier with a **PRBS** sequence produces a sinc-shaped signal centered at the carrier frequency. Several definitions of bandwidth exist: Null-to-null bandwidth (BW of main lobe), half-power bandwidth (Where the power is down by half the peak value) and the bandwidth that contains 99% of the signal power. The main lobe null-to-null bandwidth is twice the clock rate of the modulating code and is expressed as:

$$\mathrm{BW}_{ss} = 2R_c. \tag{7.6}$$

The side lobes have a null-to-null bandwidth equal to the code clock rate as shown in Fig. 7.14,

$$f_{\text{null}} = \pm n R_c, \tag{7.7}$$

where $n = 1, 2$. Each null occurs at a frequency location defined as

$$f_{\text{line}} = \pm f_0 \pm n R_c / N. \tag{7.8}$$

7.4 DSSS TRANSMITTER

The data waveform, m(t), is a bipolar NRZ signal transmitted at a rate of Rm bps. The PRBS waveform is also a bipolar NRZ line code, clocked at R_c chips per second (a chip being the time to transmit a single bit or symbol of the PRB sequence). The spread spectrum processing gain, G, is the ratio of the RF bandwidth to the information bandwidth and is the ratio of the data period T_m, to the PRBS period T_c defined as:

$$G = \frac{T_m}{T_c} = \frac{R_c}{R_m}. \tag{7.9}$$

The processing gain is increased by making the PRBS longer. Typically, a commercial DSS radio processing gain is 20 dB but it depends on the data rate and tolerates jamming power levels much stronger than the desired signal. The maximal length of the m stage serial shift register is

$$N = 2^m - 1. \tag{7.10}$$

The number of chips is $N = 2^m - 1 = 2^6 - 1 = 63$. The m-sequence code length period in seconds is

$$T_{\text{DS}} = N T_c. \tag{7.11}$$

Alternatively, (7.11) is written as

$$R_{\text{DS}} = \frac{R_c}{N}. \tag{7.12}$$

Here $R_c = 10^5$ chips/s and has an individual spectral line spacing shown in Fig. 7.15, whose value is

$$R_{\text{DS}} = \frac{10^5}{63} = 1587 \text{ Hz}. \tag{7.13}$$

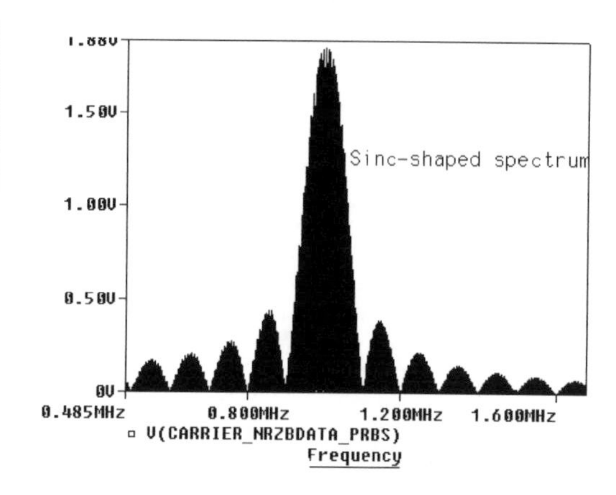

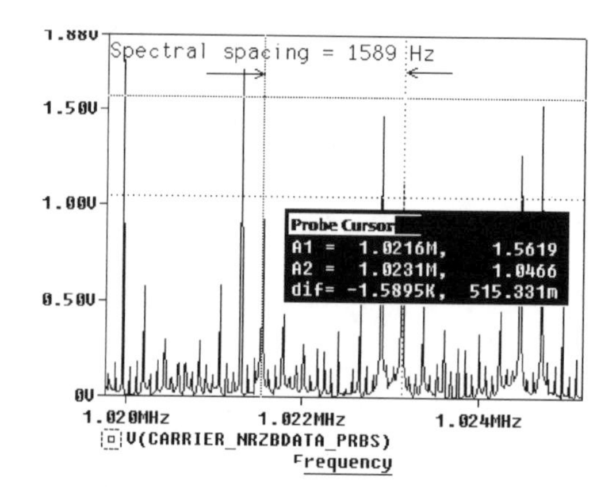

FIGURE 7.15: Spectral spacing is 1586 Hz.

The PRBS has a clock period of 10 μs (1/100 kHz), and the data period is 10ms (1/100 Hz). The length of the sequence, N, is measured from the spectrum by selecting the **FFT** icon. The difference between the spectral lines in the spectrum is an indirect measurement of the PRBS length, N. We measure the difference as 1586 Hz, thus N is 1/(fc*1586) = 63. The frequency spacing for a 100 kHz clock is

$$f = \frac{1}{NT_c} = \frac{1}{63 \times 10^{-5}} = 1586\ Hz. \tag{7.14}$$

Equation (7.14) is useful for obtaining the PRBS length when the sequence is large and difficult to measure, i.e.,

$$N = \frac{1}{fT_c} = \frac{1}{1586 \times 10^{-5}} = 63. \tag{7.15}$$

We have to replace the random data with a regular square wave to measure the spectral line spacing of 1586 Hz.

7.5 SPREAD SPECTRUM RECEIVER

Because of the PSpice evaluation limitations, it is necessary to use a **VPWL_F_RE_FOREVER** generator part as the input to the DSS receiver. This part reads in the file copied from **PROBE** screen after simulating the DSS transmitter.

From Probe, copy using ctrl C applied to the variable **V(carrier_NRZBData_PRBS)** located on the top plot on the x-axis in Fig. 7.12. This should turn the selected variable red. **CTRL C** copies the signal for pasting in WordPad©. The first line shown in Fig. 7.16 must be deleted. Failure to delete this line will produce an error message when simulated. Save the signal as "spreadsignal4.txt."

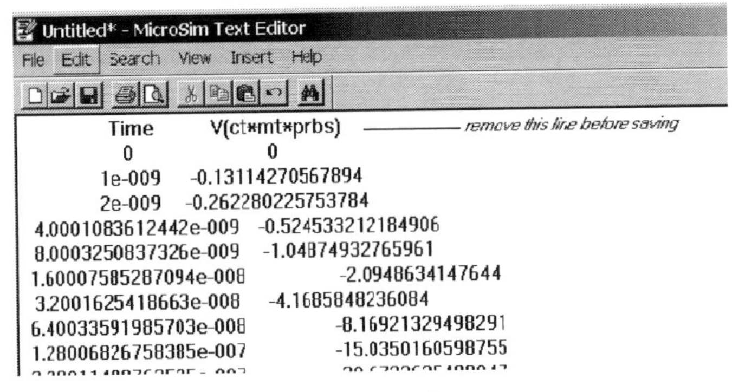

FIGURE 7.16: Pasting the output SS signal in Notepad.

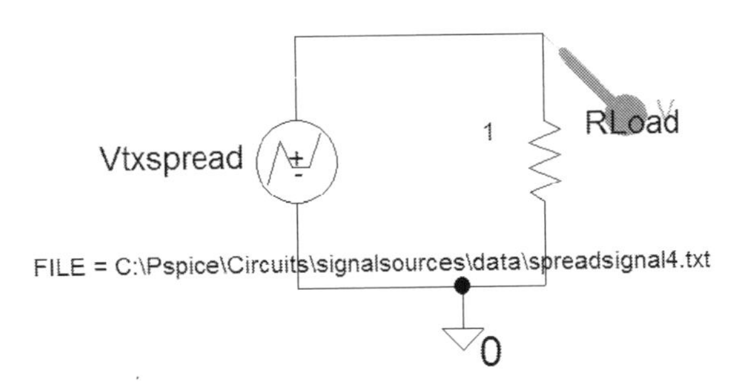

FIGURE 7.17: Testing the SS signal.

Test the saved SS signal "spreadsignal4.txt" using the schematic shown in Fig. 7.17.

The **VPWL_F_RE_FOREVER** generator parameter File is filled in by highlighting the generator, **Rclicking** and **selecting Edit Properties**. Included in the schematic is the original data generator to compare to the recovered data signal. The transmitted SS signal $m(t)$ = "spread1.txt" is applied and de-spread by multiplying it by the original **PRBS** sequence sent from the transmitter and the same carrier. Multiplying two cosine signals results in two signals, one of which has a frequency twice the frequency of the original signal message and is removed by a LPF (it introduces a delay in the recovered data signal, however). Add a comparator after the filtered output to shape the signal so that it can be compared to the original data (use if then else in an **ABM1** part for this).

The original input data signal uses the **STIM1** part whose parameters are shown in Fig. 7.11. The noise generator output is initially set to zero by making the voltage scaling factor (VSF) zero.

The recovered output signals are shown in Fig. 7.19.

Press the **FFT** icon to display the frequency spectrum shown in Fig. 7.20.

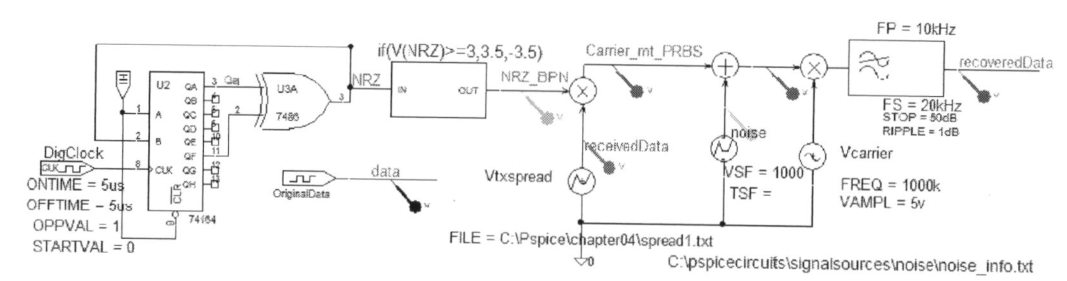

FIGURE 7.18: Spread spectrum receiver.

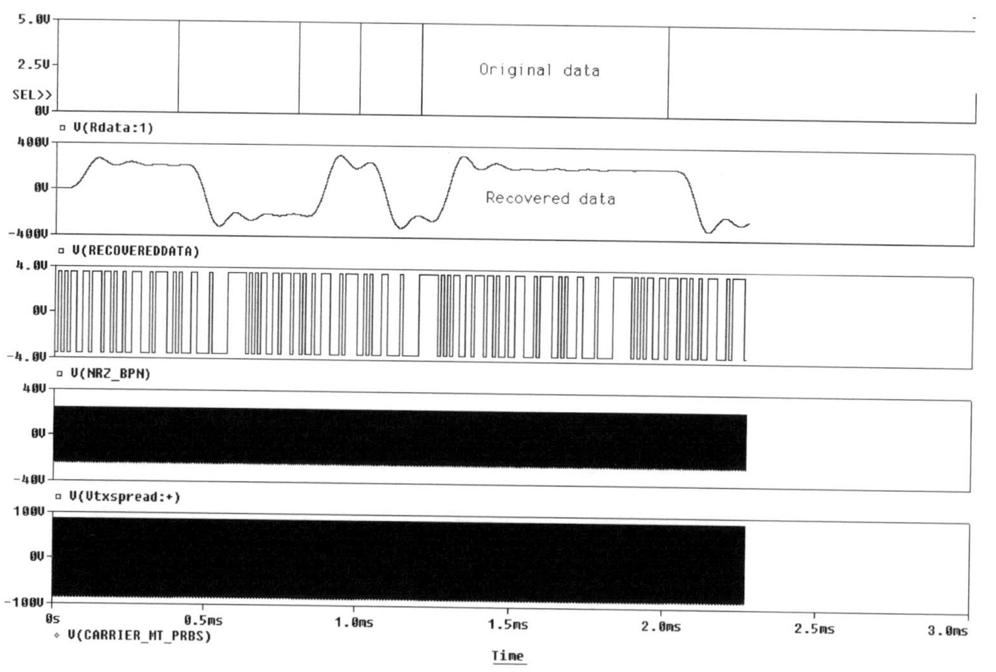

FIGURE 7.19: Receiver signals.

7.6 ADDING NOISE TO THE RECEIVED SIGNAL

One of the advantages of SS is the ability to recover the signal in the presence of noise, even if the noise is larger than the received signal. The schematic in Fig. 7.18 shows how noise is added to the received signal to simulate channel noise. Increase the VSF in order to increase the noise. In Fig. 7.18, VSF = 1000.

The effect of adding noise to the input received SS signal is shown in Fig. 7.22.

It is left as an exercise to replace the **ABM** multiplier and filter with actual parts such as the AD633 IC.

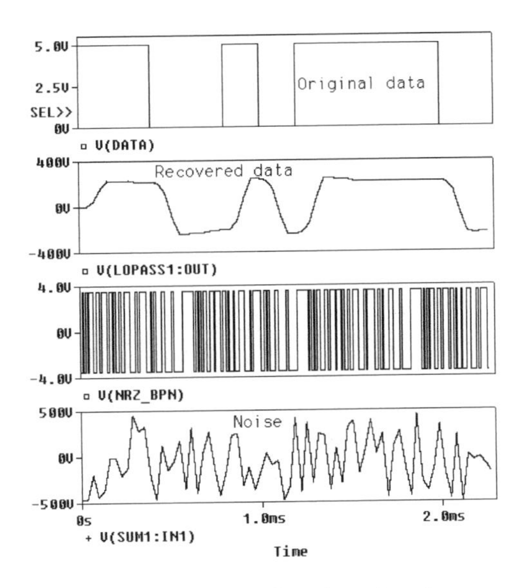

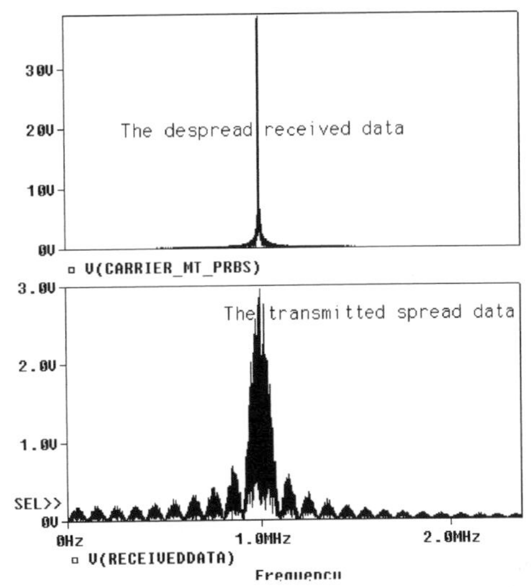

FIGURE 7.20: Spectrum of receiver signals.

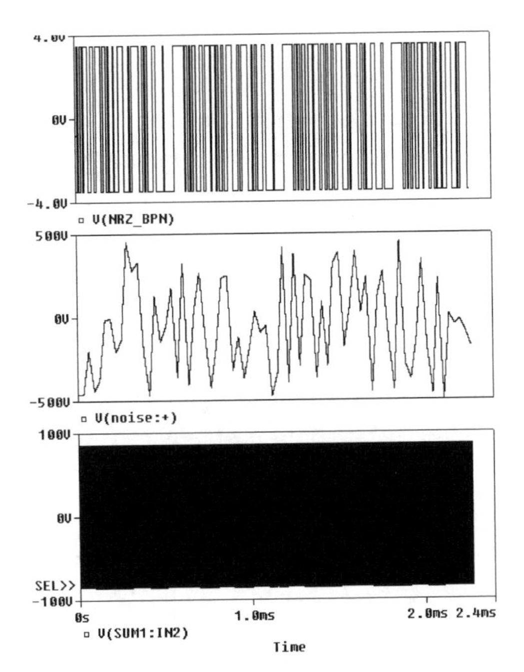

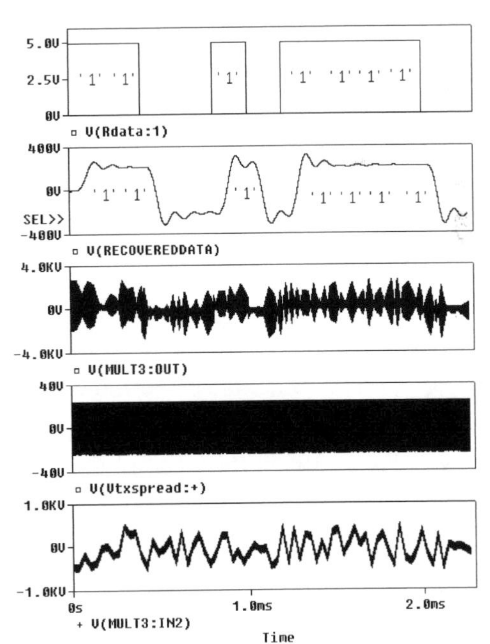

FIGURE 7.21: Recovered DSSS signals.

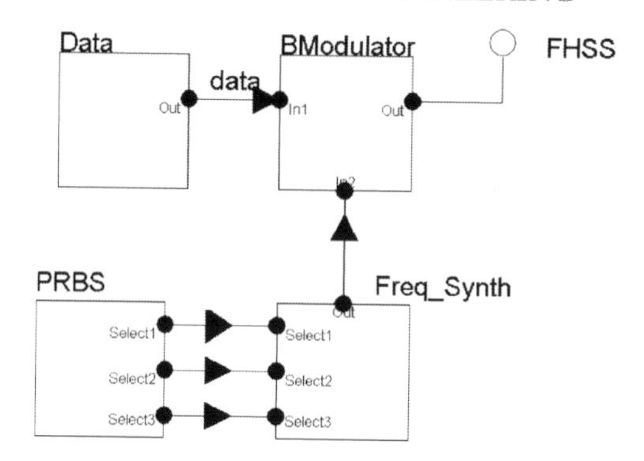

FIGURE 7.22: FHSS.

7.7 FREQUENCY-HOPPING SPREAD SPECTRUM

Frequency-hopping spread spectrum (FHSS) transmits a short burst at one frequency and then hops to another frequency thus spreading the signal over a wide spectrum. The transmitter and receiver are synchronized in order to be on the same frequency at the same time. The frequency-hopping pattern (order in which they are used and the dwell time at each frequency) are regulated. For example, the FCC requires 75 frequencies or more and with a maximum dwell time of 400ms. Data is retransmitted on a subsequent hop to another frequency if interference occurs on that frequency. All FHSS products allow users to deploy more than one channel in the same area by implementing separate channels on different, orthogonal, hopping sequences. Frequency hopping is a spread spectrum technique used in Bluetooth—a popular technique for connecting mobile devices, such as laptops to mobile phones etc.

In FHSS, the data message is broken into blocks of data and causes a random carrier frequency to be generated. The carrier frequency hops from one frequency to another frequency in a random fashion, the hopping being driven by a PRBS generator. The frequency-hopping rate is a function of the information rate, since the sequence is driven by the message data. As in DSSS, the PRBS is "known" by the transmitter and receiver before transmission starts. The FHSS spectrum is flat over the band of frequencies, unlike the (sin x/x)-shaped spectrum produced by a DSSS. Dehopping in the receiver uses an identical PRBS to that used in the transmitter and it drives the receiver local oscillator frequency synthesizer, producing the same carrier demodulating frequencies. The FHSS system in Fig. 7.22 contains a frequency synthesizer, PRBS generator, balanced modulator, and word generator.

The frequency hopping is achieved using a PLL synthesizer that contains a frequency divider that drives a multiplexer whose output is switched by three signals from a PRBS generator. Thus the signals from the divider can be switched in and out of the circuit depending

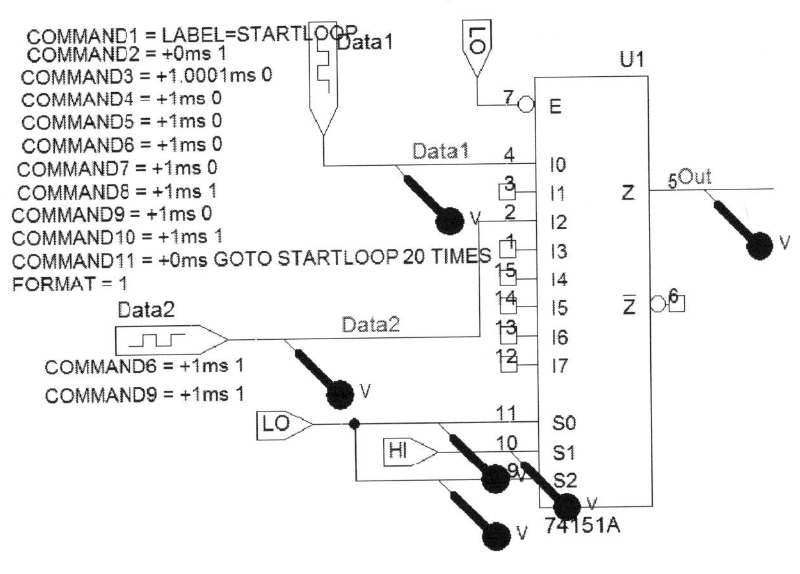

FIGURE 7.23: Multiplexer.

on the code generated by the PRBS circuit. To understand the circuit operation, we must examine and test each part of the system separately.

7.7.1 Multiplexer

Fig. 7.23 shows a multiplexer with data applied to pins 10 and 12. What we are trying to do here is to make the data "Data2" applied to pin 12, appear on the Z output, when the binary code "010" is applied to pins S0, S1, and S2. If we tied all three pins to LO, i.e., "000," then any data applied to pin 10 would be routed to the Z output.

The selection of Data2 is shown in Fig. 7.24.

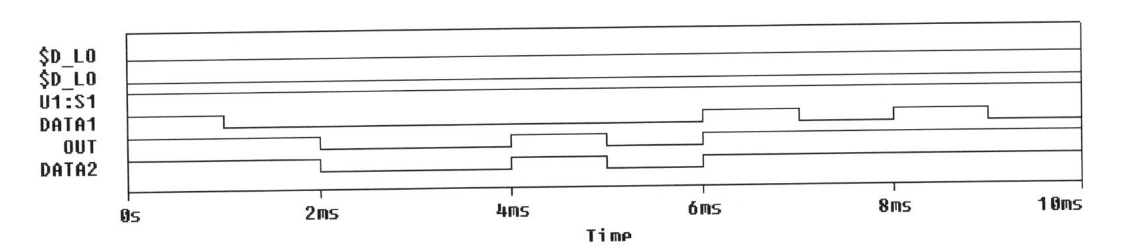

FIGURE 7.24: "010" selecting Data2.

7.7.2 PRBS

The PRBS in Fig. 7.23 outputs three signals, Select1, Select2, and Select3, that randomly select the frequency divider outputs (decade counter) and route them through the multiplexer. Insert this multiplexer into the PLL shown in Fig. 7.25. S0, S1, and S2 pins connected to the PRBS unit control the division factor N and hence the carrier frequency.

The PLL frequency synthesizer in Fig. 7.26 shows a 7490 decade counter that produces outputs f_0/N. The outputs for this IC are as follows:

- Pin 12, called QA, yields a divide by 2.
- Pin 9, called QB, yields a divide by 4.
- Pin 8, called QC, yields a divide by 8.
- Pin 11, called QD, yields a divide by 16.

The VCO is set to f_0, and has a stable frequency driving the PLL.

The output signals are shown in Fig. 7.27.

7.8 EXERCISE

1. Construct and simulate, a complete FHSS and to start you off consider Fig. 7.22 which combines the schematics shown in Fig. 7.25 and Fig. 7.26.

It is left as an exercise to test the complete system. However, since we cannot plot spectrograms in PSpice we cannot show the random carrier selection with time.

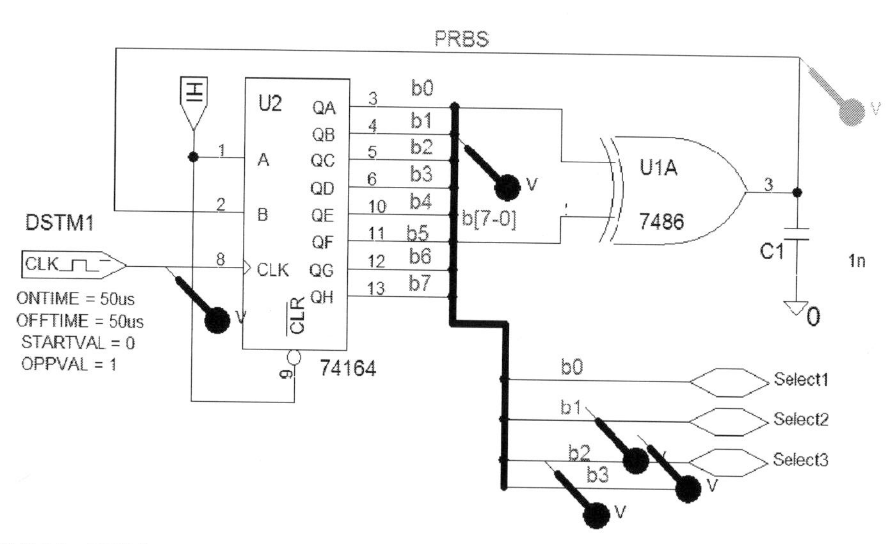

FIGURE 7.25: PRBS.

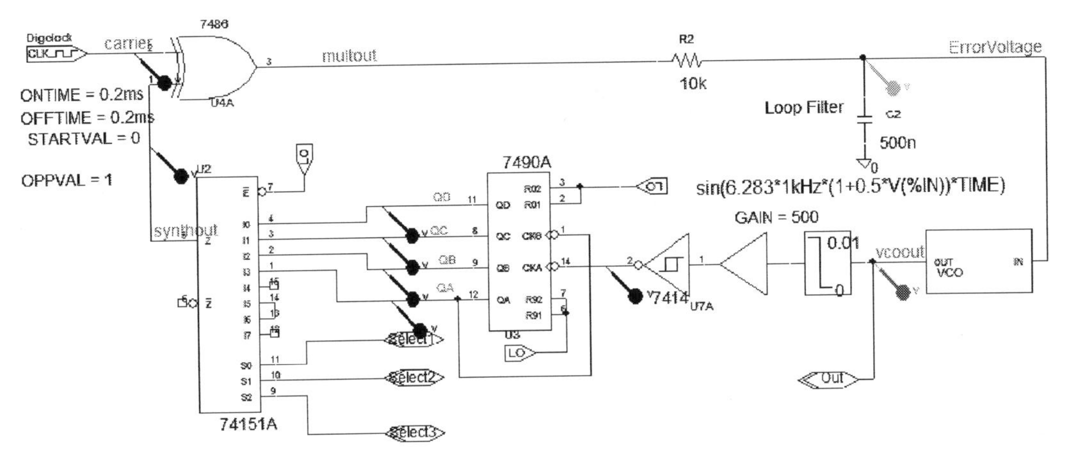

FIGURE 7.26: FHSS phase a lock loop.

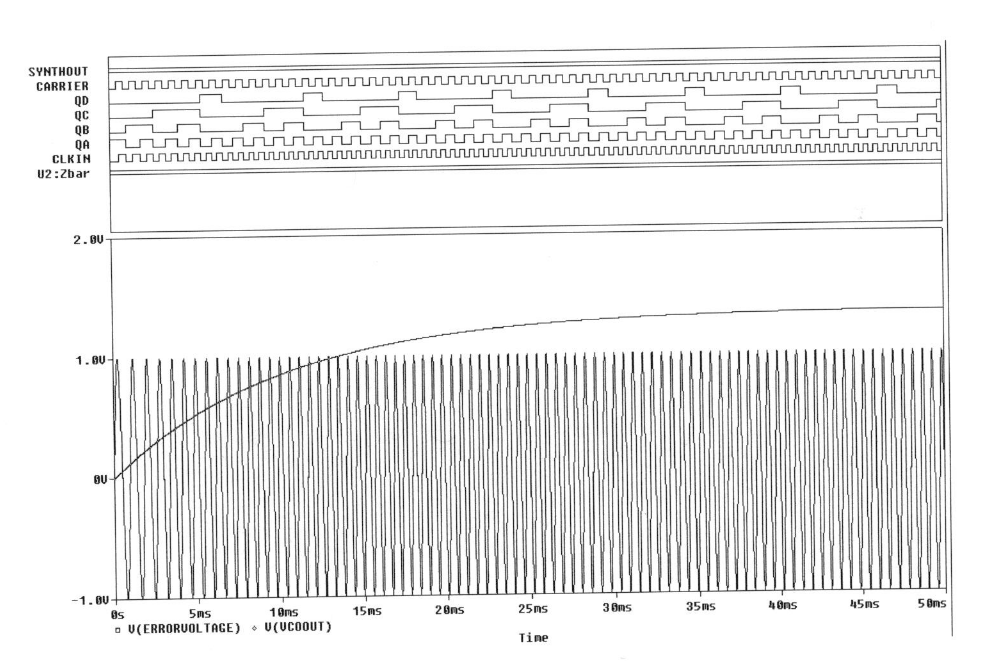

FIGURE 7.27: Frequency synthesizer waveforms.

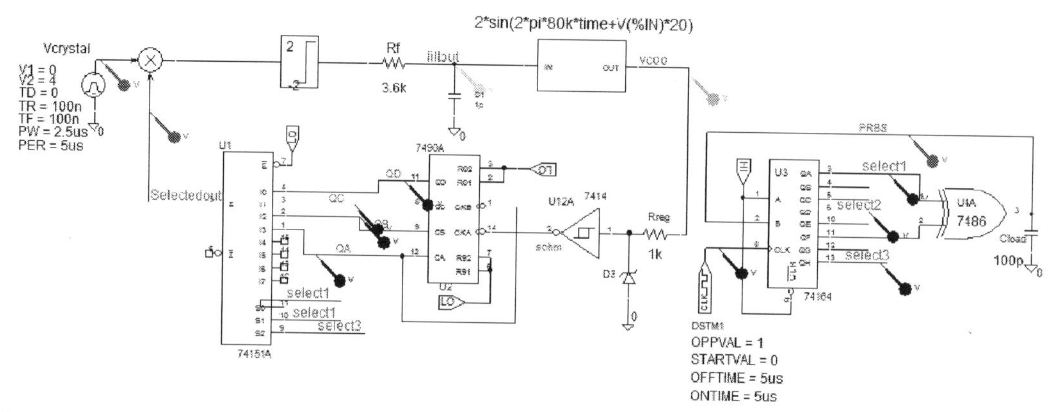

FIGURE 7.28: Frequency synthesizer driven by PRBS.

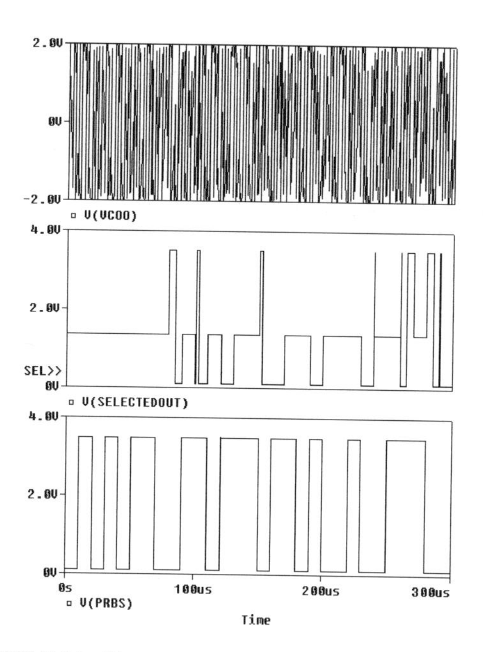

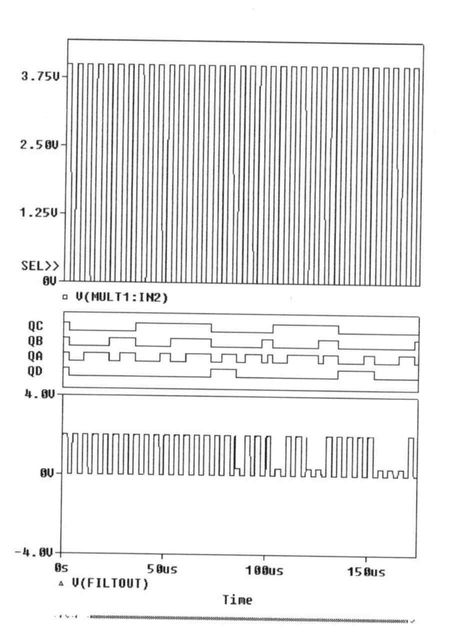

FIGURE 7.29: Frequency synthesizer signals.

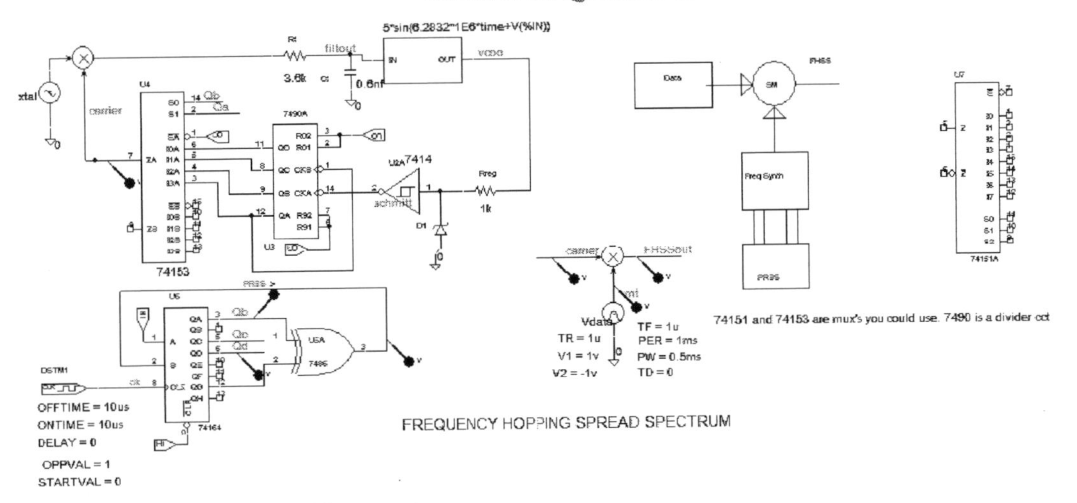

FIGURE 7.30: Possible FHSS transmitter.

Appendix A: References

BOOKS

1. Tobin P (Feb 2007) PSpice for Filters and Transmission Lines. Morgan Claypool publishers.
2. Tobin P (2007) PSpice for Digital Communications Engineering. Morgan Claypool publishers.
3. Bateman Andy (1998) Digital Communications-Design for the Real World, Prentice Hall.
4. Young P H (2003) Electronic Communications Techniques. Maxwell-MacMillan.
5. Harold Kolimbiris, Digital communication systems, Person Education.
6. Digital Signal Processing: A Practical Approach, Second Edition Emmanuel C Ifeachor and Barrie W Jervis, ISBN 0 201 59619 9.
7. Concepts In Systems and Signals by John D Sherrick.
8. Tobin Paul (Feb 2007) PSpice for Digital signal processing. Morgan Claypool publishers.
9. Tobin Paul (Feb 2007) PSpice for Circuit theory and Electric devices Morgan Claypool publishers.
10. Tobin Paul (Feb 2007) PSpice for Analog Communications Engineering Morgan Claypool publishers.

INTERNET

- http://en.wikipedia.org/wiki/Shannon-Hartley_theorem
- http://www.dspguru.com/info/faqs/multrate/resamp.htm
- http://www.analog.com/processors/resources/beginnersGuide/index.html
- http://www.arrl.org/tis/info/sdr.html
- http://www.freqdev.com/guide/dspguide.html#digdesign
- http://www.bores.com/courses/intro/program/index.htm
- http://www.dsprelated.com/showmessage/20912/1.php
- http://web.mit.edu/6.555/www/fir.html
- http://www.analog-innovations.com/
- http://www.complextoreal.com

Appendix B: Tables

TABLE B-1: Fourier Transform Table

FUNCTION	$f(t)$	FOURIER TRANSFORM
Constant	1	$\delta(f)$
Impulse	$\delta(t - t_0)$	$e^{(-j2\pi f_0)}$
Square wave pulse	T	$T\frac{\sin(\pi f T)}{\pi f T}$
Cosine	$\cos(\omega_0 t + \theta)$	$0.5e^{j\theta}\delta(f - f_0) + 0.5e^{-j\theta}\delta(f + f_0)$
Triangular signal	T	$T\left[\frac{\sin(\pi f T)}{\pi f T}\right]^2$
Delay T_d	$x(t - T_d)$	$X(f)e^{(-j2\pi f T_d)}$
Frequency translation	$x(t)e^{(j2\pi f_0 t)}$	$V(f - f_0)$

TABLE B-2: Laplace and z-Transform Table

FUNCTION	$F(t)$	LAPLACE TRANSFORM	$f(n)$	z-transform ($t = nT = n$)
Unit step	$u(t)$	$1/s$	$u(n)$	$z/z - 1$
Unit impulse	$\delta(t)$	1	$\delta(n)$	1
Unit ramp	T	$1/s^2$	N	$nz/(z-1)^2$
Polynomial	t^n	$n!/s^{n+1}$	t^n	$T^2 z(z+1)/(z-1)^2$ for $n = 2$
Decaying exponential	e^{-at}	$1/(s+a)$	$e^{-an}u(n)$	$z/z - e^{-an}$
Growing Exponential	$1/a(1-e^{-at})$	$1/(s+a)(s)$	$1/a(1-e^{-an})$	$z(1-e^{-an})/a(z-1)(z-e^{-an})$
Sine	$\sin(\omega t)$	$\omega/(s^2+\omega^2)$	$\sin(n\theta)u(n)$	$(z \sin n\theta)/(z^2 - 2z \sin n\theta + 1)$
Cosine	$\cos(\omega t)$	$s/(s^2+\omega^2)$	$\cos(n\theta)u(n)$	$\dfrac{z(z - \cos n\theta)}{z^2 - 2z\cos n\theta + 1}$
Damped sine	$e^{-at}\sin(\omega t)$	$\omega/((s+a)^2+\omega^2)$	$e^{-an}\sin(n\theta)$	$\dfrac{ze^{-an}\sin(n\theta)}{z^2 - 2ze^{-an}\cos n\theta + e^{-2an}}$
Damped cosine	$e^{-at}\cos(\omega t)$	$(s+a)/((s+a)^2+\omega^2)$	$e^{-an}\cos(n\theta)$	$\dfrac{z^2 - ze^{-an}\cos(n\theta)}{z^2 - 2ze^{-an}\cos n\theta + e^{-2an}}$
Delay	$f(t-k)$	e^{-sk}	$f(n-k)$	z^{-k}

TABLE B-3: Bessel Functions

BETA	J(0)	J(1)	J(2)	J(3)	J(4)	J(5)	J(6)	J(7)	J(8)	J(9)
0	1	0	0	0	0	0	0	0	0	0
0.25	0.9844	0.124	0	0	0	0	0	0	0	0
0.5	0.9385	0.2423	0.0306	0	0	0	0	0	0	0
0.75	0.8642	0.3492	0.0671	0	0	0	0	0	0	0
1	0.7652	0.4401	0.1149	0.0196	0	0	0	0	0	0
1.25	0.6459	0.5106	0.1711	0.0369	0	0	0	0	0	0
1.5	0.5118	0.5579	0.2321	0.061	0.0118	0	0	0	0	0
1.75	0.369	0.5802	0.294	0.0919	0.0209	0	0	0	0	0
2	0.2239	0.5767	0.3528	0.1289	0.034	0	0	0	0	0
2.25	0.0827	0.5484	0.4047	0.1711	0.0515	0.0121	0	0	0	0
2.4	0.0025	0.5202	0.431	0.1981	0.0643	0.0162	0	0	0	0
2.5	−0.0484	0.4971	0.4461	0.2166	0.0738	0.0195	0	0	0	0
2.75	−0.1641	0.426	0.4739	0.2634	0.1007	0.0297	0	0	0	0
3	−0.2601	0.3391	0.4861	0.3091	0.132	0.043	0.0114	0	0	0
3.5	−0.3801	0.1374	0.4586	0.3868	0.2044	0.0804	0.0254	0	0	0
4	−0.3971	−0.066	0.3641	0.4302	0.2811	0.1321	0.0491	0.0152	0	0
4.5	−0.3205	−0.2311	0.2178	0.4247	0.3484	0.1947	0.0843	0.03	0.0091	0
4.75	−0.2551	−0.2892	0.1334	0.4015	0.3738	0.228	0.1063	0.0405	0.0131	0
5	−0.1776	−0.3276	0.0466	0.3648	0.3912	0.2611	0.131	0.0534	0.0184	0
5.5	−0.0068	−0.3414	−0.1173	0.2561	0.3967	0.3209	0.1868	0.0866	0.0337	0.0113

TABLE B-4: Useful Trigonometric Identities

FUNCTION	EXPANSION
$\sin A \sin B$	$0.5\cos(A-B) - 0.5\cos(A+B)$
$\cos A \cos B$	$0.5\cos(A-B) + 0.5\cos(A+B)$
$\sin A \cos B$	$0.5\sin(A-B) + 0.5\sin(A+B)$
$e^{\pm j\theta}$	$\cos\theta \pm j\sin\theta$
$\cos\theta$	$(e^{j\theta} + e^{-j\theta})/2$
$\sin\theta$	$(e^{j\theta} - e^{-j\theta})/2j$
$1 = \cos^2\theta + \sin^2\theta$	
$\cos 2\theta = \cos^2\theta - \sin^2\theta$	

Index

Author Biography

Paul Tobin graduated from Kevin Street College of Technology (now the Dublin Institute of Technology) with honours in electronic engineering and went to work for the Irish National Telecommunications company. Here, he was involved in redesigning the analogue junction network replacing cables with PCM systems over optical fibres. He gave a paper on the design of this new digital junction network to the Institute of Engineers of Ireland in 1982 and was awarded a Smith testimonial for one of the best papers that year. Having taught part-time courses in telecommunications systems in Kevin Street, he was invited to apply for a full-time lecture post. He accepted and started lecturing full time in 1983. Over the last twenty years he has given courses in telecommunications, digital signal processing and circuit theory.

He graduated with honours in 1998 having completed a taught MSc in various DSP topics and a project using the Wavelet Transform and neural networks to classify EEG (brain waves) associated with different mental tasks. He has been a 'guest professor' in the Institut Universitaire de Technologie (IUT), Bethune, France for the past four years giving courses in PSpice simulation topics. He wrote an unpublished book on PSpice but was persuaded by Joel Claypool (of Morgan and Claypool Publishers) at an engineering conference in Puerto Rico (July 2006), to break it into five PSpice books. One of the books introduces a novel way of teaching DSP using PSpice. There are over 500 worked examples in the five books covering a range of topics with sufficient theory and simulation results from basic circuit theory right up to advanced communication principles. Most of these worked examples have been thoroughly 'student tested' by Irish and International students and should mean little or no errors but alas... He is married to Marie and has four grown sons and his hobbies include playing modern jazz on double bass and piano but grew up playing G-banjo and guitar. His other hobby is flying and obtained a private pilots license (PPL) in the early 80's.

Made in the USA
San Bernardino, CA
26 March 2018